An Introduction to Stochastic
Processes in Physics

An Introduction to Stochastic Processes in Physics

Containing *"On the Theory of Brownian Motion"* by Paul Langevin, translated by Anthony Gythiel

DON S. LEMONS

THE JOHNS HOPKINS UNIVERSITY PRESS

BALTIMORE AND LONDON

The Johns Hopkins University Press
2715 North Charles Street
Baltimore, Maryland 21218-4363
www.press.jhu.edu

Library of Congress Cataloging-in-Publication Data

Lemons, Don S. (Don Stephen), 1949–
   An introduction to stochastic processes in physics / by Don S. Lemons
   p. cm.
   Includes bibliographical references and index.
   ISBN 0-8018-6866-1 (alk. paper) – ISBN 0-8018-6867-X (pbk. : alk. paper)
   1. Stochastic processes. 2. Mathematical physics. I. Langevin, Paul, 1872–1946. II. Title.
QC20.7.S8 L45 2001
530.15′923–dc21   2001046459

A catalog record for this book is available from the British Library

For Allison, Nathan, and Micah

Contents

Preface and Acknowledgments

Physicists have abandoned determinism as a fundamental description of reality. The most precise physical laws we have are quantum mechanical, and the principle of quantum uncertainty limits our ability to predict, with arbitrary precision, the future state of even the simplest imaginable system. However, scientists began developing probabilistic, that is, stochastic, models of natural phenomena long before quantum mechanics was discovered in the 1920s. Classical uncertainty preceded quantum uncertainty because, unlike the latter, the former is rooted in easily recognized human conditions. We are too small and the universe too large and too interrelated for thoroughly deterministic thinking.

For whatever reason—fundamental physical indeterminism, human finitude, or both—there is much we don't know. And what we do know is tinged with uncertainty. Baseballs and hydrogen atoms behave, to a greater or lesser degree, unpredictably. Uncertainties attend their initial conditions and their dynamical evolution. This also is true of every artificial device, natural system, and physics experiment.

Nevertheless, physics and engineering curriculums routinely invoke precise initial conditions and the existence of deterministic physical laws that turn these conditions into equally precise predictions. Students spend many hours in introductory courses solving Newton's laws of motion for the time evolution of projectiles, oscillators, circuits, and charged particles before they encounter probabilistic concepts in their study of quantum phenomena. Of course, deterministic models are useful, and, possibly, the double presumption of physical determinism and superhuman knowledge simplifies the learning process. But uncertainties are always there. Too often these uncertainties are ignored and their study delayed or omitted altogether.

An Introduction to Stochastic Processes in Physics revisits elementary and foundational problems in classical physics and reformulates them in the language of random variables. Well-characterized random variables quantify uncertainty and tell us what can be known of the unknown. A random variable is defined by the variety of numbers it can assume and the probability with which each number is assumed. The number of dots showing face up on a die is a random variable. A die can assume an integer value 1 through 6, and, if unbiased and honestly rolled, it is reasonable to suppose that any particular side will come up one time out of six in the long run, that is, with a probability of 1/6.

This work builds directly upon early twentieth-century explanations of the "peculiar character in the motions of the particles of pollen in water," as described in the early nineteenth century by the British cleric and biologist Robert Brown. Paul Langevin, in 1908, was the first to apply Newton's second law to a "Brownian particle," on which the total force included a random component. Albert Einstein had, three years earlier than Langevin, quantified Brownian motion with different methods, but we adopt Langevin's approach because it builds most directly on Newtonian dynamics and on concepts familiar from elementary physics. Indeed, Langevin claimed his method was "infinitely more simple" than Einstein's. In 1943 Subrahmanyan Chandrasekhar was able to solve a number of important dynamical problems in terms of probabilistically defined random variables that evolved according to Langevin's version of $F = ma$. However, his famous review article, "Stochastic Problems in Physics and Astronomy" (Chandrasekhar 1943) is too advanced for students approaching the subject for the first time.

This book is designed for those students. The theory is developed in steps, new methods are tried on old problems, and the range of applications extends only to the dynamics of those systems that, in the deterministic limit, are described by linear differential equations. A minimal set of required mathematical concepts is developed: statistical independence, expected values, the algebra of normal variables, the central limit theorem, and Wiener and Ornstein-Uhlenbeck processes. Problems append each chapter. I wanted the book to be one I could give my own students and say, "Here, study this book. Then we will do some interesting research."

Writing a book is a lonely enterprise. For this reason I am especially grateful to those who aided and supported me throughout the process. Ten years ago Rick Shanahan introduced me to both the concept of and literature on stochastic processes and so saved me from foolishly trying to reinvent the field. Subsequently, I learned much of what I know about stochastic processes from Daniel Gillespie's excellent book (Gillespie 1992). Until his recent, untimely death, Michael Jones of Los Alamos National Laboratory was a valued partner in exploring new applications of stochastic processes. Memory eternal, Mike! A sabbatical leave from Bethel College allowed me to concentrate on writing during the 1999–2000 academic year. Brian Albright, Bill Daughton, Chris Graber, Bob Harrington, Ed Staneck, and Don Quiring made valuable comments on various parts of the typescript. Willis Overholt helped with the figures. More general encouragement came from Reuben Hersh, Arnold Wedel, and Anthony Gythiel. I am grateful for all of these friends.

**An Introduction to Stochastic
Processes in Physics**

1

Random Variables

1.1 Random and Sure Variables

A quantity that, under given conditions, can assume different values is a *random variable*. It matters not whether the random variation is intrinsic and unavoidable or an artifact of our ignorance. Physicists can sometimes ignore the randomness of variables. Social scientists seldom have this luxury. The total number of "heads" in ten coin flips is a random variable. So also is the range of a projectile. Fire a rubber ball through a hard plastic tube with a small quantity of hair spray for propellant. Even when you are careful to keep the tube at a constant elevation, to inject the same quantity of propellant, and to keep all conditions constant, the projectile lands at noticeably different places in several trials. One can imagine a number of causes of this variation: different initial orientations of a not-exactly-spherical ball, slightly variable amounts of propellant, and breeziness at the top of the trajectory. In this as well as in similar cases we distinguish between *systematic error* and *random variation*. The former can, in principle, be understood and quantified and thereby controlled or eliminated. Truly random sources of variation cannot be associated with determinate physical causes and are often too small to be directly observed. Yet, unnoticeably small and unknown random influences can have noticeably large effects.

A random variable is conceptually distinct from a *certain* or *sure variable*. A sure variable is, by definition, exactly determined by given conditions. Newton expressed his second law of motion in terms of sure variables. Discussions of sure variables are necessarily cast in terms of concepts from the ivory tower of physics: perfect vacuums, frictionless pulleys, point charges, and exact initial conditions. The distance an object falls from rest, in a perfect vacuum, when constantly accelerating for a definite period of time is a sure variable.

Just as it is helpful to distinguish notationally between scalars and vectors, it is also helpful to distinguish notationally between random and sure variables. As is customary, we denote random variables by uppercase letters near the end of the alphabet, for example, $V, W, X, Y,$ and Z, while we denote sure variables by lowercase letters, for example, $a, b, c, x,$ and y. The time evolution of a random variable is called a *random* or *stochastic process*. Thus $X(t)$ denotes a stochastic process. The time evolution of a sure variable is called a *deterministic process* and could be denoted by $x(t)$. Sure variables and deterministic processes are familiar mathematical objects. Yet, in a sense, they are idealizations of random variables and processes.

Modeling a physical process with sure instead of random variables involves an assumption—sometimes an unexamined assumption. How do we know, for instance, that the time evolution of a moon of Jupiter is a deterministic process while the time evolution of a small grain of pollen suspended in water is a random process? What about the phase of a harmonic oscillator or the charge on a capacitor? Are these sure or random variables? How do we choose between these two modeling assumptions?

That all physical variables and processes are essentially random is the more general of the two viewpoints. After all, a sure variable can be considered a special kind of random variable—one whose range of random variation is zero. Thus, we adopt as a working hypothesis that all physical variables and processes are random ones. The details of a theory of random variables and processes will tell us under what special conditions sure variables and deterministic processes are good approximations. We develop such a theory in the chapters that follow.

1.2 Assigning Probabilities

A random variable X is completely specified by the range of values x it can assume and the probability $P(x)$ with which each is assumed. That is to say, the probabilities $P(x)$ that $X = x$ for all possible values of x tell us everything there is to know about the random variable X. But how do we assign a number to "the probability that $X = x$"? There are at least two distinct answers to this question—two interpretations of the word *probability* and, consequently, two interpretations of the phrase *random variable*. Both interpretations have been with us since around 1660, when the fundamental laws of mathematical probability were first discovered (Hacking 1975).

Consider a coin toss and associate a random variable X with each possible outcome. For instance, when the coin lands heads up, assign $X = 1$, and when the coin lands tails up, $X = 0$. To determine the probability $P(1)$ of a heads-up outcome, one could flip the coin many times under identical conditions and form the ratio of the number of heads to the total number of coin flips. Call that ratio $f(1)$. According to the *statistical* or *frequency* interpretation of probability, the ratio $f(1)$ approaches the probability $P(1)$ in the limit of an indefinitely large number of flips. One virtue of the frequency interpretation is that it suggests a direct way of measuring or, at least, estimating the probability of a random outcome. The English statistician J. E. Kerrich so estimated $P(1)$ while interned in Denmark during World War II (Kerrich 1946). He flipped a coin 10,000 times and found that heads landed uppermost in 5067 "spins." Therefore, $P(1) \approx f(1) = 0.5067$—at least for Kerrich's coin and method of flipping.

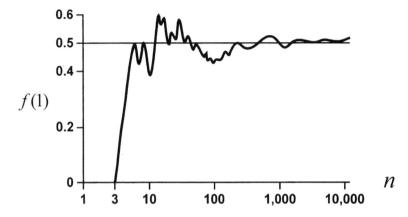

Figure 1.1. Frequency of heads, $f(1)$, versus number of flips, n. Replotted, from Kerrich 1946.

Kerrich's was not the first heroic frequency measurement. In 1850 the Swiss astronomer Wolf rolled one white and one red die 20,000 times, kept track of the results, and determined the frequency of each outcome (Bulmer 1967). Also, a certain nineteenth-century English biologist Weldon also rolled twelve dice 26,306 times and recorded the number of 5s and 6s (Fry 1928).

That actual events can't be repeated ad infinitum doesn't invalidate the frequency interpretation of probability any more than the impossibility of a perfect vacuum invalidates the law of free fall. Both are idealizations that make a claim about what happens in a series of experiments as an unattainable condition is more and more closely approached. In particular, the frequency interpretation claims that fluctuations in $f(1)$ around $P(1)$ become smaller and smaller as the number of coin flips becomes larger and larger. Because Kerrich's data, in fact, has this feature (see figure 1.1), his coin flip can be considered a random event with its defining probabilities, $P(1)$ and $P(0)$, equal to the limiting values of $f(1)$ and $f(0)$.

An alternative method of determining $P(1)$ is to inspect the coin and, if you can find no reason why one side should be favored over the other, simply assert that $P(1) = P(0) = 1/2$. This method of assigning probabilities is typical of the so-called *degree of belief* or *inductive* interpretation of probability. According to this view, a probability quantifies the truth-value of a proposition. In physics we are primarily concerned with propositions of the form $X = x$. In assigning an inductive probability $P(X = x)$, or simply $P(x)$, to the proposition $X = x$, we make a statement about the degree to which $X = x$ is believable. Of course, if they are to be useful, inductive probabilities should not be assigned haphazardly but rather should reflect the available evidence and change when that evidence changes. In this account probability theory extends deductive logic to cases involving partial implication—thus the name *inductive probability*. Observe that inductive probabilities can be assigned to any outcome, whether repeatable or not.

The *principle of indifference*, devised by Pierre Simon Laplace (1749–1827), is one procedure for assigning inductive probabilities. According to this principle, which was invoked above in asserting that $P(1) = P(0) = 1/2$, one should assign equal probabilities to different outcomes if there is no reason to favor one outcome over any other. Thus, given a seemingly unbiased six-sided die, the inductive probability of any one side coming up is $1/6$. The *principle of equal a priori probability*, that a dynamical system in equilibrium has an equal probability of occupying each of its allowed states, is simply Laplace's principle of indifference in the context of statistical mechanics. The *principle of maximum entropy* is another procedure for assigning inductive probabilities. While a good method for assigning inductive probabilities isn't always obvious, this is more a technical problem to be overcome than a limitation of the concept.

That the laws of probability are the same under both of these interpretations explains, in part, why the practice of probabilistic physics is much less controversial than its interpretation, just as the practice of quantum physics is much less controversial than its interpretation. For this reason one might be tempted to embrace a mathematical agnosticism and be concerned only with the rules that probabilities obey and not at all with their meaning. But a scientist or engineer needs some interpretation of probability, if only to know when and to what the theory applies.

The best interpretation of probability is still an open question. But probability as quantifying a degree of belief seems the most inclusive of the possibilities. After all, one's degree of belief could reflect an in-principle indeterminism or an ignorance born of human finitude or both. Frequency data is not required for assigning probabilities, but when available it could and should inform one's degree of belief. Nevertheless, the particular random variables we study also make sense when their associated probabilities are interpreted strictly as limits of frequencies.

1.3 The Meaning of Independence

Suppose two unbiased dice are rolled. If the fact that one shows a "5" doesn't change the probability that the other also shows a "5," the two outcomes are said to be *statistically independent*, or simply *independent*. When the two outcomes are independent and the dice unbiased, the probability that both dice will show a "5" is the product $(1/6)(1/6) = 1/36$. While statistical independence is the rule among dicing outcomes, the random variables natural to classical physics are often statistically dependent. For instance, one usually expects the location X of a particle to depend in some way upon its velocity V.

Let's formalize the concept of statistical independence. If realization of the outcome $X = x$ does not change the probability $P(y)$ that outcome $Y = y$ obtains and vice-versa, the outcomes $X = x$ and $Y = y$ are statistically independent and the probability that they occur jointly $P(x\&y)$ is the product $P(x)P(y)$, that is,

$$P(x\&y) = P(x)P(y). \qquad (1.3.1)$$

When condition (1.3.1) obtains for all possible realizations x and y, the random variables X and Y are said to be statistically independent. If, on the other hand, the realization $X = x$ does change the probability $P(y)$ that $Y = y$ or vice-versa, then

$$P(x\&y) \neq P(x)P(y) \qquad (1.3.2)$$

and the random variables X and Y are *statistically dependent*.

The distinction between independent and dependent random variables is crucial. In the next chapter we construct a numerical measure of statistical dependence. And in subsequent chapters we will, on several occasions, exploit special sets of explicitly independent and dependent random variables.

Problems

1.1. Coin Flipping. Produce a graph of the frequency of heads $f(1)$ versus the number of coin flips n. Use data obtained from

a. flipping a coin 100 times,
b. pooling your coin flip data with that of others, or
c. numerically accessing an appropriate random number generator 10,000 times.

Do fluctuations in $f(1)$ obtained via method a, b, and c diminish, as do those in figure 1.1, as more data is obtained?

1.2 Independent Failure Modes. A system consists of n separate components, each one of which fails independently of the others with probability P_i where $i = 1 \ldots n$. Since each component must either fail or not fail, the probability that the ith component does not fail is $1 - P_i$.

a. Suppose the components are connected in parallel so that the failure of all the components is necessary to cause the system to fail. What is the probability the system fails? What is the probability the system functions?

b. Suppose the components are connected in series so that the failure of any one component causes the system to fail. What is the probability the system fails? (Hint: First, find the probability that all components function.)

2

Expected Values

2.1 Moments

The *expected value* of a random variable X is a function that turns the probabilities $P(x)$ into a sure variable called the *mean* of X. The mean is the one number that best characterizes the possible values of a random variable. We denote the mean of X variously by mean$\{X\}$ and $\langle X \rangle$ and define it by

$$\langle X \rangle = \sum_i x_i P(x_i) \tag{2.1.1}$$

where the sum is over all possible realizations x_i of X. Thus, the mean number of dots showing on an unbiased die is $(1+2+3+4+5+6)/6 = 3.5$. The square of a random variable is also a random variable. If the possible realizations of X are the numbers 1, 2, 3, 4, 5, and 6, then their squares, 1, 4, 9, 16, 25, and 36, are the possible realizations of X^2. In fact, any algebraic function $f(X)$ of a random variable X is also a random variable. The expected value of the random variable $f(X)$ is denoted by $\langle f(X) \rangle$ and defined by

$$\langle f(X) \rangle = \sum_i f(x_i) P(x_i). \tag{2.1.2}$$

The mean $\langle X \rangle$ parameterizes the random variable X, but so also do all the *moments* $\langle X^n \rangle$ $(n = 0, 1, 2, \ldots)$ and *moments about the mean* $\langle (X - \langle X \rangle)^n \rangle$. The operation by which a random variable X is turned into one of its moments is one way of asking X to reveal its properties, or *parameters*. Among the moments about the mean,

$$\langle (X - \langle X \rangle)^0 \rangle = \langle 1 \rangle$$

$$= \sum_i P(x)$$

$$= 1 \tag{2.1.3}$$

simply recovers the fact that probabilities are normalized. And

$$\langle (X - \langle X \rangle)^1 \rangle = \sum_i (x_1 - \langle X \rangle) P(x_i)$$

$$= \sum_i x_i P(x_i) - \langle X \rangle \sum_i P(x_i)$$

$$= \langle X \rangle - \langle X \rangle \langle 1 \rangle$$

$$= 0 \tag{2.1.4}$$

follows from normalization (2.1.3) and the definition of the mean (2.1.1). Higher order moments (with $n \geq 2$) describe other properties of X. For instance, the second moment about the mean or the *variance* of X, denoted by var$\{X\}$ and defined by

$$\text{var}\{X\} = \langle (X - \langle X \rangle)^2 \rangle, \tag{2.1.5}$$

quantifies the variability, or mean squared deviation, of X from its mean $\langle X \rangle$. The linearity of the expected value operator $\langle \rangle$ (see section 2.2) ensures that (2.1.5) reduces to

$$\text{var}\{X\} = \langle X^2 - 2X\langle X \rangle + \langle X \rangle^2 \rangle$$

$$= \langle X^2 \rangle - 2\langle X \rangle^2 + \langle X \rangle^2$$

$$= \langle X^2 \rangle - \langle X \rangle^2. \tag{2.1.6}$$

The mean and variance are sometimes denoted by the Greek letters μ and σ^2, respectively, and $\sqrt{\sigma^2} = \sigma$ is called the *standard deviation* of X. The third moment about the mean enters into the definition of *skewness*,

$$\text{skewness}\{X\} = \frac{\langle (X - \mu)^3 \rangle}{\sigma^3}, \tag{2.1.7}$$

and the fourth moment into the *kurtosis*,

$$\text{kurtosis}\{X\} = \frac{\langle (X - \mu)^4 \rangle}{\sigma^4}. \tag{2.1.8}$$

The skewness and kurtosis are dimensionless shape parameters. The former quantifies the asymmetry of X around its mean, while the latter is a measure of the degree to which a given variance σ^2 is accompanied by realizations of X close to (relatively small kurtosis) and far from (large kurtosis) $\mu \pm \sigma$. Highly peaked and long-tailed probability functions have large kurtosis; broad, squat ones have small kurtosis. See Problem 2.1, *Dice Parameters*, for practice in calculating parameters.

2.2 Mean Sum Theorem

The sum of two random variables is also a random variable. As one might expect, the probabilities and parameters describing $X + Y$ are combinations of the probabilities and parameters describing X and Y separately. The expected value of a sum is defined in terms of the joint probability $P(x_i \& y_i)$ that both $X = x_i$ and $Y = y_i$, that is, by

$$\langle X + Y \rangle = \sum_i \sum_j (x_i + y_j) P(x_i \& y_j). \tag{2.2.1}$$

That

$$\begin{aligned} \langle X + Y \rangle &= \sum_i x_i \sum_j P(x_i \& y_j) + \sum_j y_j \sum_i P(x_i \& y_j) \\ &= \sum_i x_i P(x_i) + \sum_j y_j P(y_j) \\ &= \langle X \rangle + \langle Y \rangle \end{aligned} \tag{2.2.2}$$

follows from (2.2.1) and the laws of probability. For this reason, the expected value brackets $\langle \ \rangle$ can be distributed through each term of a sum. In purely verbal terms: the mean of a sum is the sum of the means. An obvious generalization of (2.2.2) expressing the complete linearity of the operator $\langle \ \rangle$ is

$$\langle aX + bY \rangle = a\langle X \rangle + b\langle Y \rangle, \tag{2.2.3}$$

where a and b are arbitrary sure values.

We will have occasions to consider multiple-term sums of random variables such as

$$X = X_1 + X_2 + \cdots + X_n \tag{2.2.4}$$

where n is very large or even indefinitely large. For instance, a particle's total displacement X in a time interval is the sum of the particle's successive displacements X_i (with $i = 1, 2, \ldots n$) in successive subintervals. Because the mean of a sum is the sum of the means,

$$\langle X \rangle = \langle X_1 \rangle + \langle X_2 \rangle + \cdots + \langle X_n \rangle, \tag{2.2.5}$$

or, equivalently,

$$\text{mean} \left\{ \sum_{i=1}^{n} X_i \right\} = \sum_{i=1}^{n} \text{mean}\{X_i\}. \tag{2.2.6}$$

We call (2.2.5) and (2.2.6) the *mean sum theorem*.

2.3 Variance Sum Theorem

The moments of the product XY are not so easily expressed in terms of the separate moments of X and Y. Only in the special case that X and Y are statistically independent can we make statements similar in form to the mean sum theorem. In general,

$$\langle XY \rangle = \sum_i \sum_j x_i y_j P(x_i \,\&\, y_j). \tag{2.3.1}$$

But when X and Y are statistically independent, $P(x_i \,\&\, y_j) = P(x_i)P(y_j)$ and equation (2.3.1) reduces to

$$\langle XY \rangle = \sum_i x_i P(x_i) \sum_j y_j P(y_y), \tag{2.3.2}$$

which is equivalent to

$$\langle XY \rangle = \langle X \rangle \langle Y \rangle, \tag{2.3.3}$$

that is, the mean of a product is the product of the means. Statistical independence also ensures that

$$\langle X^n Y^m \rangle = \langle X^n \rangle \langle Y^m \rangle \tag{2.3.4}$$

for any n and m. If it happens that $\langle X^n Y^m \rangle = \langle X^n \rangle \langle Y^m \rangle$ for some but not all n and m, then X and Y are not statistically independent.

When the random variables X and Y are dependent, we can't count on $\langle XY \rangle$ factoring into $\langle X \rangle \langle Y \rangle$. The *covariance*

$$
\begin{aligned}
\operatorname{cov}\{X, Y\} &= \langle (X - \langle X \rangle)(Y - \langle Y \rangle) \rangle \\
&= \langle [XY - \langle X \rangle Y - X \langle Y \rangle + \langle X \rangle \langle Y \rangle] \rangle \\
&= \langle XY \rangle - \langle X \rangle \langle Y \rangle
\end{aligned}
\tag{2.3.5}
$$

and the *correlation coefficient*

$$\operatorname{cor}\{X, Y\} = \frac{\operatorname{cov}\{X, Y\}}{\sqrt{\operatorname{var}\{X\}\operatorname{var}\{Y\}}} \tag{2.3.6}$$

are measures of the statistical dependence of X and Y. The correlation coefficient establishes a dimensionless scale of dependence and independence such that $-1 \le \operatorname{cor}\{X, Y\} \le 1$. When X and Y are *completely correlated*, so that X and Y realize the same values on the same occasions, we say that $X = Y$. In this case $\operatorname{cov}\{X, Y\} = \operatorname{var}\{X\} = \operatorname{var}\{Y\}$ and $\operatorname{cor}\{X, Y\} = 1$. When X and

Y are completely *anticorrelated*, so that $X = -Y$, $\text{cor}\{X, Y\} = -1$. When X and Y are statistically independent, so that $\langle XY \rangle = \langle X \rangle \langle Y \rangle$, $\text{cov}\{X, Y\} = 0$ and $\text{cor}\{X, Y\} = 0$. See Problem 2.2, *Perfect Linear Correlation*.

We exploit the concept of covariance in simplifying the expression for the variance of a sum of two random variables. We call

$$
\begin{aligned}
\text{var}\{X+Y\} &= \langle (X+Y-\langle X+Y \rangle)^2 \rangle \\
&= \langle (X-\langle X \rangle)^2 \rangle + \langle (Y-\langle Y \rangle)^2 \rangle + 2\langle (X-\langle X \rangle)(Y-\langle Y \rangle) \rangle \\
&= \langle (X-\langle X \rangle)^2 \rangle + \langle (Y-\langle Y \rangle)^2 \rangle + 2(\langle XY \rangle - \langle X \rangle \langle Y \rangle) \\
&= \text{var}\{X\} + \text{var}\{Y\} + 2\,\text{cov}\{X, Y\} \qquad (2.3.7)
\end{aligned}
$$

the *variance sum theorem*. It reduces to the *variance sum theorem for independent addends*

$$
\text{var}\{X + Y\} = \text{var}\{X\} + \text{var}\{Y\} \qquad (2.3.8)
$$

only when X and Y are statistically independent. Repeated application of (2.3.8) to a sum of n statistically independent random variables leads to

$$
\text{var}\left\{ \sum_{i=1}^{N} X_i \right\} = \sum_{i=1}^{N} \text{var}\{X_i\}. \qquad (2.3.9)
$$

Thus, the variance of a sum of independent variables is the sum of their variances.

For instance, suppose we wish to express the mean and variance of the area A of a rectangular plot of land in terms of the mean and variance of its length L and width W. If L and W are statistically independent, $\langle LW \rangle = \langle L \rangle \langle W \rangle$ and $\langle L^2 W^2 \rangle = \langle L^2 \rangle \langle W^2 \rangle$. Then

$$
\begin{aligned}
\text{mean}\{A\} &= \langle LW \rangle \\
&= \langle L \rangle \langle W \rangle \qquad (2.3.10)
\end{aligned}
$$

and

$$
\begin{aligned}
\text{var}\{A\} &= \langle A^2 \rangle - \langle A \rangle^2 \\
&= \langle L^2 W^2 \rangle - \langle LW \rangle^2 \\
&= \langle L^2 \rangle \langle W^2 \rangle - \langle L \rangle^2 \langle W \rangle^2. \qquad (2.3.11)
\end{aligned}
$$

Given that $\langle L^2 \rangle = \text{var}\{L\} + \langle L \rangle^2$ and $\langle W^2 \rangle = \text{var}\{W\} + \langle W \rangle^2$, equations

(2.3.10) and (2.3.11) achieve the desired result. For other applications of the mean and variance sum theorems, see Problem 2.3, *Resistors in Series*, and Problem 2.4, *Density Fluctuations*.

2.4 Combining Measurements

How do we combine different measurements of the same random quantity? Suppose, for instance, I use a meter stick to measure the width of the table on which I write. My procedure produces a realization x_1 of a random variable X_1. The variable X_1 is random because the table sides are not perfectly parallel, its ends are not well defined, I must visually interpolate between the smallest marks on the rule to get the last digit, my eyesight is not so good, nor is my hand perfectly steady, and the meter stick is not really rigid. Now, suppose I tilt the table surface and measure its angle of incline to the horizontal, time a marble rolling across the table width, measure the marble's radius, and from this data and the local acceleration of gravity compute the table width. For similar reasons, this number x_2 is also the realization of a random variable X_2. Finally, I use a laser interferometer and electronically count fringes as the interferometer mirror is moved across the table. This procedure results in a number x_3 that is the realization of a third random variable X_3. Among the three numbers x_1, x_2, and x_3, which is the best measurement of the table width? Assuming I avoid systematic errors (for example, I don't use a meter stick whose end has been cut off), then

$$\langle X_1 \rangle = \langle X_2 \rangle = \langle X_3 \rangle \qquad (2.4.1)$$

because each procedure measures the same quantity—the table width. However, the different procedures accumulate random error in different amounts, and these will be reflected in their different variances. If the interferometer measurement x_3 is the least prone to random error, then $\text{var}\{X_3\} < \text{var}\{X_1\}$ and $\text{var}\{X_3\} < \text{var}\{X_2\}$. In this sense, x_3 is the best measurement.

But is x_3 any better than the arithmetical average

$$\bar{x} = \frac{1}{3}(x_1 + x_2 + x_3)? \qquad (2.4.2)$$

Before the mid-eighteenth century, scientists were reluctant to average measurements that were produced in substantially different ways. They feared the most precise measurement, in this case x_3, would be "contaminated" by those of lesser precision, in this case x_2 and x_3—that "errors would multiply, not compensate" (Stigler 1986). The issue is easily resolved given the insight that the average $\bar{x}$ is a particular realization of the random variable

$$\bar{X} = \frac{1}{3}(X_1 + X_2 + X_3). \qquad (2.4.3)$$

Now $\bar{X}$ has a mean,

$$\langle \bar{X} \rangle = \langle X_1 \rangle = \langle X_2 \rangle = \langle X_3 \rangle, \qquad (2.4.4)$$

and a variance,

$$\begin{aligned} \text{var}\{\bar{X}\} &= \langle (\bar{X})^2 \rangle - \langle \bar{X} \rangle^2 \\ &= \frac{1}{9}[\text{var}\{X_1\} + \text{var}\{X_2\} + \text{var}\{X_3\}]. \end{aligned} \qquad (2.4.5)$$

In deriving the latter we have assumed that X_1, X_2, and X_3 are statistically independent and employed the variance sum theorem for independent addends (2.3.9). If $\text{var}\{X_3\} < \text{var}\{\bar{X}\}$, then x_3 is a better measurement than $\bar{x}$, and x_3 would be contaminated if averaged together with x_1 and x_2. If, on the other hand, $\text{var}\{\bar{X}\} < \text{var}\{X_1\}$, $\text{var}\{\bar{X}\} < \text{var}\{X_2\}$, and $\text{var}\{\bar{X}\} < \text{var}\{X_3\}$, then the average $\bar{x}$ is better than any one of the values from which it is composed. In this case the errors in x_1, x_2, and x_3 compensate for each other in the average $\bar{x}$. Either ordering is possible.

In general, although not always, the more terms included in the average, the better *statistic*, or estimator, it becomes. Suppose we devise n different, independent ways of making the same measurement. The random variable representing the average measurement is

$$\bar{X} = \frac{(X_1 + X_2 + \cdots + X_n)}{n}, \qquad (2.4.6)$$

and the variance of the average is

$$\text{var}\{\bar{X}\} = \frac{\sum_{i=1}^{n} \text{var}\{X_i\}}{n^2}. \qquad (2.4.7)$$

Because the numerator of the right-hand side of (2.4.7) increases (roughly) with n and the denominator increases with n^2, the variance of the average $\bar{X}$ decreases with increasing n as $1/n$. Thus, averaging is generally a good idea.

Averaging is, in fact, always helpful if all the measurements are made in the same way. Jacob Bernoulli put it this way in 1731: "For even the most stupid of men, by some instinct of nature, by himself and without any instruction (which is a remarkable thing), is convinced that the more observations have been made, the less danger there is of wandering from one's goal" (Stigler 1986). Hence, if all the measurements are made in the same way,

$$\text{var}\{X_1\} = \text{var}\{X_2\} = \ldots \text{var}\{X_n\} \qquad (2.4.8)$$

and given (2.4.7), the variance of the average is

$$\text{var}\{\bar{X}\} = \frac{\text{var}\{X_1\}}{n}. \qquad (2.4.9)$$

Figure 2.1. Resistors in series.

Furthermore, the standard deviation of the average is

$$\text{std}\{\bar{X}\} = \sqrt{\text{var}\{\bar{X}\}}$$

$$= \frac{\text{std}\{X_1\}}{\sqrt{n}}. \tag{2.4.10}$$

Thus, the more terms included in the average, the smaller the standard deviation of the average. As n becomes indefinitely large, $\bar{X}$ approaches a random variable whose variance vanishes, that is, $\bar{X}$ approaches the sure value $\langle \bar{X} \rangle$.
The standard deviation divided by the mean,

$$\frac{\text{std}\{\bar{X}\}}{\langle \bar{X} \rangle} = \frac{1}{\sqrt{n}} \frac{\text{std}\{X_1\}}{\langle X_1 \rangle}, \tag{2.4.11}$$

measures the *precision* of a particular measurement and is called the *coefficient of variation*. The smaller the coefficient of variation, the more likely is each realization of $\bar{X}$ close to $\langle \bar{X} \rangle$. Problem 2.4, *Density Fluctuations*, applies this mathematics in another context.

Problems

2.1. Dice Parameters.   An unbiased die realizes each of its values, 1, 2, 3, 4, 5, and 6, with equal probability $1/6$. Find the mean, variance, standard deviation, skewness, and kurtosis of the random variable X so defined.

2.2. Perfect Linear Correlation.   Two random variables X and Y are related by $Y = mX + b$. This means that every realization x_i of X is related to a realization y_i of Y by $y_i = mx_i + b$ where m and b are sure variables. Prove that $\text{cor}\{X, Y\} = m/\sqrt{m^2} = \text{sgn}\{m\}$ where $\text{sgn}\{m\}$ is the sign of m.

2.3. Resistors in Series.   You are given a box of n carbon resistors (see figure 2.1). On each the manufacturer has color-coded a nominal resistance, which we understand to be a mean$\{R_i\}$, and a dimensionless "tolerance" or "precision" t_i whose definition we take to be

$$t_i = \frac{\sqrt{\text{var}\{R_i\}}}{\text{mean}\{R_i\}} \times 100\%$$

where $i = 1 \ldots n$. Assume the resistances R_i are statistically independent random variables.

a. Write expressions for the mean, variance, and tolerance of the total resistance R of a series combination of n identically defined resistors in terms of the mean$\{R_i\}$ and tolerance t_i of one resistor.

b. Suppose the box contains 10 nominally 5-Ohm resistors, each with a 20% tolerance. Calculate the mean, variance, and tolerance of the resistance of their series combination. Is the tolerance of this combination less than the tolerance of the separate resistors? It should be.

2.4. Density Fluctuations. The molecular number density $\rho = N/V$ of a gas contained in a small open region of volume V within a larger closed volume V_0 fluctuates as the number of molecules N in V changes. To quantify fluctuations in the density ρ, let the larger volume V_0 contain exactly N_0 molecules (figure 2.2). The number N can be considered a sum of statistically independent auxiliary "indicator" random variables X_i, defined so that $X_i = 1$ when molecule i is within volume V and $X_i = 0$ when it is not. Then,

$$N = \sum_{i=1}^{N_o} X_i.$$

Assume, as is reasonable, that when the gas is in equilibrium,

$$P(X_i = 1) = \frac{V}{V_o}$$

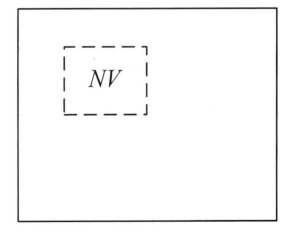

$$N_0 V_0$$

Figure 2.2. The number of molecules N within a small open volume V is a random variable. The total number of molecules N_0 within the larger closed volume V_0 is a sure variable.

and

$$P(X_i = 0) = \frac{V_o - V}{V_o}$$

for all i.

a. Compute mean$\{X_i\}$ and var$\{X_i\}$ in terms of the constants V_o, and V.

b. Determine mean$\{N\}$, var$\{N\}$, and the coefficient of variation

$$\sqrt{\text{var}\{N\}}/\,\text{mean}\{N\}$$

in terms of N_o, V_o, and V.

3

Random Steps

3.1 Brownian Motion Described

We are ready to use our knowledge of how random variables add and multiply to model the simplest of all physical processes—a single particle at rest. If at one instant a particle occupies a definite position and has zero velocity, it will, according to Newton's first law of motion, continue to occupy the same position as long as no forces act on it. Consider, though, whether this deterministic (and boring) picture can ever be a precise description of any real object. Even when great care is taken to isolate the particle, there are always air molecules around to nudge it one way or the other.

If the particle is very small ($\leq 50 \times 10^{-6}$m), the net effect of these nudges can be observed in an optical microscope. These *Brownian motions* are so called after the Scottish naturalist and cleric Robert Brown (1773–1858), who investigated the phenomenon in 1827. (That Jan IngenHousz [1730–1799)], a Dutch-born biologist, observed and described Brownian motion even earlier, in 1785, is just one of many illustrations of *Stigler's Law of Eponymy*—which states that no discovery is named after its original discoverer.) When looking through a microscope at grains of pollen suspended in water, Brown noticed that a group of grains always disperses and that individual grains move around continuously and irregularly. Brown originally thought that he had discovered the irreducible elements of a vitality common to all life forms. However, upon systematically observing these irregular motions in pollen from live and dead plants, in pieces of other parts of plants, in pieces of animal tissue, in fossilized wood, in ground window glass, various metals, granite, volcanic ash, siliceous crystals, and even in a fragment of the Sphinx, he gave up that hypothesis.

We now know that Brownian motion is a consequence of the atomic theory of matter. When a particle is suspended in any fluid media (air as well as water), the atoms or molecules composing the fluid hit the particle from different directions in unequal numbers during any given interval. While the human eye cannot distinguish the effect of individual molecular impacts, it can observe the net motion caused by many impacts over a period of time.

3.2 Brownian Motion Modeled

Let's model Brownian motion as a sum of independent random displacements. Imagine the Brownian particle starts at the origin $x = 0$ and is free to move in either direction along the x-axis. The net effect of many individual molecular impacts is to displace the particle a random amount X_i in each interval of duration Δt. Assume each displacement X_i realizes one of two possibilities, $X_i = +\Delta x$ or $X_i = -\Delta x$, with equal probabilities ($\frac{1}{2}$) and that the various X_i are statistically independent. After n such intervals the net displacement X is

$$X = X_1 + X_2 + \cdots + X_n. \tag{3.2.1}$$

This is the *random step* or *random walk* model of Brownian motion. According to the model,

$$\langle X_1 \rangle = \langle X_2 \rangle = \ldots \langle X_n \rangle = 0 \tag{3.2.2}$$

since $\langle X_i \rangle = (1/2)(+\Delta x) + (1/2)(-\Delta x) = 0$ for each $i = 1, 2, \ldots n$. Therefore, the mean sum theorem yields

$$\begin{aligned} \langle X \rangle &= \langle X_1 \rangle + \langle X_2 \rangle + \cdots + \langle X_n \rangle \\ &= 0, \end{aligned} \tag{3.2.3}$$

that is, while any single Brownian particle may drift from its starting point, the mean of the displacement $\langle X \rangle$ maintains its initial (zero) value. Now,

$$\mathrm{var}\{X_1\} = \mathrm{var}\{X_2\} = \ldots \mathrm{var}\{X_n\} = \Delta x^2 \tag{3.2.4}$$

since

$$\begin{aligned} \mathrm{var}\{X_i\} &= \langle X_i^2 \rangle - \langle X_i \rangle^2 \\ &= \langle X_i^2 \rangle \\ &= \left(\frac{1}{2}\right)(+\Delta x)^2 + \left(\frac{1}{2}\right)(-\Delta x)^2 \\ &= \Delta x^2 \end{aligned} \tag{3.2.5}$$

for each $i = 1, 2, \ldots n$. For this reason, and because the X_i are statistically independent, the variance sum theorem yields

$$\begin{aligned} \langle X^2 \rangle &= \sum_{i=1}^{n} \mathrm{var}\{X_i\} \\ &= n\Delta x^2. \end{aligned} \tag{3.2.6}$$

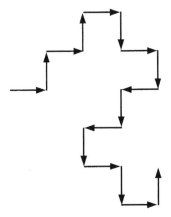

Figure 3.1. Random walk in two dimensions realized by taking alternate steps along the vertical and horizontal axes and determining the step polarity (left/right and up/down) with a coin flip.

Since the total duration of the walk is $t = n\Delta t$, equation (3.2.6) is equivalent to

$$\langle X^2 \rangle = \left(\frac{\Delta x^2}{\Delta t} \right) t. \qquad (3.2.7)$$

This equation expresses the signature property of Brownian motion: the variance $\langle X^2 \rangle$ of the net displacement X is proportional to the time t during which that displacement is made.

It is easy to generalize the one-dimensional random walk in several ways. For instance, figure 3.1 shows the effect of taking alternate displacements in different perpendicular directions and so creating Brownian motion in a plane. See also Problem 3.1 *Two-Dimensional Random Walk*. One can also suppose that either the probabilities or the step sizes are different in different directions. See, for instance, Problems 3.2, *Random Walk with Hesitation*, and 3.3, *Multistep Walk*.

3.3 Critique and Prospect

In spite of its attractions, the random step process is deficient as a physical model of Brownian motion. One deficiency is that the variance of the total displacement, as described in equation (3.2.7), seems to depend separately upon the arbitrary magnitudes Δx and Δt through the ratio $(\Delta x^2/\Delta t)$. Unless $(\Delta x^2/\Delta t)$ is itself a physically meaningful constant, the properties of the total displacement X will depend on the fineness with which it is analyzed into subincrements. That $(\Delta x^2/\Delta t)$ is, indeed, a characteristic constant—equal to twice the *diffusion constant*—will, in chapter 6, be shown to follow from the requirement of continuity, but in the present oversimplified account this claim remains unmotivated. Another difficulty with the random step model of Brownian motion is that it lacks an obvious connection to Newton's second law. Why shouldn't the integrated second law,

$$V(t) = V(0) + \frac{1}{M} \int_0^t F(t') \, dt', \qquad (3.3.1)$$

apply even when the individual impulses $\int_{t_i}^{t_i + \Delta t} F(t') \, dt'$ composing the total impulse $\int_0^t F(t') \, dt'$ are delivered randomly? In such case we might attempt to express the right-hand side of (3.3.1) as a sum of N independent, random impulses per unit mass, each with vanishing mean and a finite variance equal to, say, Δv^2, having units of speed squared. This strategy leads to

$$\langle V^2 \rangle = \left(\frac{\Delta v^2}{\Delta t} \right) t, \qquad (3.3.2)$$

an absurd result because a kinetic energy $M \langle V^2 \rangle / 2$ cannot grow without bound. We shall see that Brownian motion can, in fact, be made consistent with Newton's second law, but first some new concepts are required.

Problems

3.1. Two-Dimensional Random Walk.

a. Produce a realization of a two-dimensional random walk with the algorithm described in the caption of figure 3.1. Use either 30 coin flips or, a numerical random number generator with a large ($n \geq 100$) number of steps n.

b. Plot $X^2 + Y^2$ versus n for the realization chosen above.

3.2. Random Walk with Hesitation.
Suppose that in each interval Δt there are three equally probable outcomes: particle displaces to the left a distance Δx, particle displaces to the right a distance Δx, or particle hesitates and stays where it is. Show that the standard deviation of the net displacement X after n time intervals, each of duration Δt, is $\sqrt{\langle X^2 \rangle} = \Delta x \sqrt{2n/3}$.

3.3. Multistep Walk.
Let the independent displacements X_i of an n-step random walk be identically distributed so that mean$\{X_1\}$ = mean$\{X_2\}$ = ... mean$\{X_n\}$ = μ and var$\{X_1\}$ = var$\{X_2\}$ = ... var$\{X_n\}$ = σ^2. The net displacement is given by $X = X_1 + X_2 + \cdots + X_n$.

a. Find mean$\{X\}$, var$\{X\}$, and $\langle X^2 \rangle$ as a function of n.

b. A steady wind blows the Brownian particle, causing its steps to the right to be larger than those to the left. That is, the two possible outcomes of each step are $X_1 = \Delta x_r$ and $X_2 = -\Delta x_l$ where $\Delta x_r > \Delta x_l > 0$. Assume the probability of a step to the right is the same as the probability of a step to the left. Find mean$\{X\}$, var$\{X\}$, and $\langle X^2 \rangle$ after n steps.

3.4. Autocorrelation. According to the random step model of Brownian motion, the particle position is, after n random steps, given by

$$X(n) = \sum_{i=1}^{n} X_i$$

where the X_i are independent displacements with $\langle X_i \rangle = 0$ and $\langle X_i^2 \rangle = \Delta x^2$ for all i. Of course, after m random steps (with $m \leq n$), the particle position is $X(m)$. In general, $X(n)$ and $X(m)$ are different random variables.

a. Find cov$\{X(n), X(m)\}$.
b. Find cor$\{X(n), X(m)\}$.
c. Show that $X(n)$ and $X(m)$ become completely uncorrelated as $m/n \to 0$ and completely correlated as $m/n \to 1$. The quantity cov$\{X(n), X(m)\}$ is sometimes referred to as an *autocovariance* and cor$\{X(n), X(m)\}$ as an *autocorrelation* because they compare the same process variable at different times.

3.5. Frequency of Heads. Suppose the number of heads N in n coin flips is given by

$$N = \sum_{i=1}^{n} X_i,$$

where $X_i = 1$ means that the ith flip has turned up heads and $X_i = 0$ that it has turned up tails. Assume these two outcomes are equally probable.

a. Find mean$\{X_i\}$ and var$\{X_i\}$.
b. Find mean$\{N\}$ and var$\{N\}$.
c. Find mean$\{N/n\}$ and var$\{N/n\}$.
d. Is the answer to part c consistent with the behavior of the frequency of heads $f(1) = N/n$ in figure 1.1 (on page 3)?

4

Continuous Random Variables

4.1 Probability Densities

In order to describe the position of a Brownian particle more realistically we require a language that allows its net displacement X to realize values lying within a continuous range. The classical physical variables whose time evolution we wish to model (positions, velocities, currents, charges, etc.) are of this kind. Therefore, in place of the probability $P(x)$ that $X = x$ we require a probability $p(x)\,dx$ that X falls within the interval $(x, x + dx)$. The function $p(x)$ is a *probability density*. Because probabilities are dimensionless, the probability density $p(x)$ has the same units as $1/x$. A continuous random variable X is completely defined by its probability density $p(x)$.

The probability $p(x)\,dx$ obeys the same rules as does $P(x)$, even if these must be formulated somewhat differently. For instance, probability densities are normalized,

$$\int_{-\infty}^{\infty} p(x)\,dx = 1, \qquad (4.1.1)$$

because some value $X = x$ must be realized. The probability density $p(x)$ must be non-negative. Also, two random variables X and Y are statistically independent if and only if their joint, $p(x\ \&\ y)$, and individual, $p(x)$ and $p(y)$, probability densities are related by

$$p(x\ \&\ y) = p(x)p(y). \qquad (4.1.2)$$

The expected value $\langle X \rangle$ of a continuous random variable X is given by

$$\langle X \rangle = \int_{-\infty}^{\infty} x p(x)\,dx. \qquad (4.1.3)$$

We will have occasions to adopt specific probability densities $p(x)$ as modeling assumptions. Among them are those defining the uniform, normal, and Cauchy random variables. Also see Problems 4.3, *Exponential Random Variable*, and 4.4, *Poisson Random Variable*.

4.2 Uniform, Normal, and Cauchy Densities

The *uniform random variable* $U(m, a)$ is defined by the probability density

$$p(x) = \frac{1}{2a} \quad \text{when } (m - a) \le x \le (m + a);$$

$$p(x) = 0 \quad \text{otherwise.} \tag{4.2.1}$$

See figure 4.1. We say that $U(m, a)$ is a uniform random variable with *center* m and *half-width* a. Note that this density is normalized, so that

$$\begin{aligned}
\text{mean}\{U(m, a)\} &= \langle U(m, a) \rangle \\
&= \int_{-\infty}^{\infty} x p(x)\, dx \\
&= \frac{1}{2a} \int_{m-a}^{m+a} x\, dx \\
&= m
\end{aligned} \tag{4.2.2}$$

and that

$$\begin{aligned}
\text{var}\{U(m, a)\} &= \langle (U(m, a) - m)^2 \rangle \\
&= \int_{-\infty}^{\infty} (x - m)^2 p(x)\, dx
\end{aligned}$$

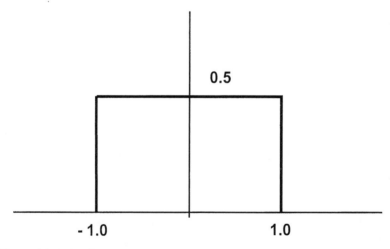

Figure 4.1. Probability density defining a uniform random variable $U(0, 1)$ with center 0 and half-width 1.

$$= \frac{1}{2a} \int_{m-a}^{m+a} (x - m)^2 \, dx$$

$$= \frac{a^2}{3}. \tag{4.2.3}$$

Other moments about the mean are given by

$$\langle (X - \langle X \rangle)^n \rangle = \frac{1}{2a} \int_{m-a}^{m+a} (x - m)^n \, dx$$

$$= \frac{a^{n+1} - (-a)^{n+1}}{2a(n+1)}. \tag{4.2.4}$$

Thus, $\langle (X - \langle X \rangle)^n \rangle = 0$ when n is odd.

$U(m, a)$ represents a quantity about which we know nothing except that it falls within a certain range $(m - a, m + a)$. Numbers taken from analog and digital measuring devices are of this kind. For instance, the "reading" 3.2 is actually the random number $U(3.2, 0.05)$ because its last significant digit, 2, is the result of taking a number originally found with uniform probability density somewhere within the interval $(3.15, 3.25)$ and rounding it up or down. Digital computers also employ particular realizations of uniform random numbers.

The *normal random variable* $N(m, a^2)$, defined by the probability density

$$p(x) = \frac{\exp[-(x - m)^2 / 2a^2]}{\sqrt{2\pi a^2}} \qquad -\infty \le x \le \infty \tag{4.2.5}$$

and illustrated in Figure 4.2, is especially useful in random process theory. The parameters m and a^2 are, by design, the mean and variance of $N(m, a^2)$. The moments of $N(m, a^2)$ about its mean are given by

$$\langle (N(m, a^2) - m)^n \rangle = \frac{1}{\sqrt{2\pi a^2}} \int_{-\infty}^{\infty} (x - m)^n \exp \left[\frac{-(x - m)^2}{2a^2} \right] dx$$

$$= 1 \cdot 3 \cdot 5 \ldots (n - 1) \cdot a^n \quad \text{for even } n, \text{ and}$$

$$= 0 \qquad\qquad\qquad \text{for odd } n. \tag{4.2.6}$$

From (4.2.6) we find that

$$kurtosis\{N(m, a^2)\} = \frac{\langle (N(m, a^2) - m)^4 \rangle}{\langle (N(m, a^2) - m)^2 \rangle^2}$$

$$= \frac{3a^4}{a^4}$$

$$= 3. \tag{4.2.7}$$

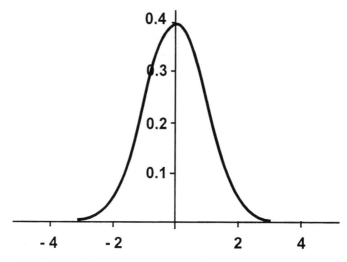

Figure 4.2. Probability density defining a normal random variable $N(0, 1)$ with mean 0 and variance 1.

The kurtosis of a normal variable is taken as a standard of comparison. When the kurtosis of a random variable is greater than 3, its probability density is said to *leptokurtic* (after the Greek word λεπτος, for "thin"), and when it is less than 3, the probability density is *platykurtic* (after πλατυξ meaning "broad"). For instance, the uniform density, which has a kurtosis of 1.8, is platykurtic. The normal probability density function is also known as a *Gaussian curve* or a *bell curve*, and, when molecular speeds are the independent variable, a *Maxwellian*.

All random variables must obey the normalization law $\langle X^0 \rangle = 1$, but the other moments don't even have to exist. In fact, the *Cauchy random variable* $C(m, a)$, with center m and half-width a, defined by

$$p(x) = \frac{(a/\pi)}{(x - m)^2 + a^2} \qquad -\infty \leq x \leq \infty \qquad (4.2.8)$$

appears to have infinite even moments. Actually, neither the odd nor the even moments of $C(m, a)$ exist in the usual sense of an improper integral with limits tending to $\pm\infty$. Thus $C(m, a)$ is maximally leptokurtic, with a thin peak and long tails (see figure 4.3). Still, $C(m, a)$ can represent physically motivated probability densities (see Problem 4.1, *Single-Slit Diffraction*). Spectral line shapes, called *Lorentzians*, also assume this form. The Cauchy density takes its name from the French mathematician Augustin Cauchy (1789–1857).

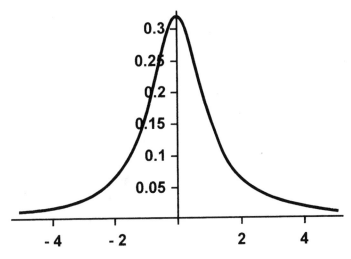

Figure 4.3. Probability density defining the Cauchy random variable $C(0, 1)$, with center 0 and half-width 1.

Figure 4.4 compares the uniform, normal, and Cauchy densities. In the limit $a \to 0$ of vanishing variance or half-width, each of the three random variables $U(m, a)$, $N(m, a^2)$, and $C(m, a)$ collapses to its mean or center m. So we can write

$$m = U(m, 0)$$
$$= N(m, 0)$$
$$= C(m, 0). \tag{4.2.9}$$

4.3 Moment-Generating Functions

Moment-generating functions are a convenient way to calculate the moments of a random variable. By definition, the *moment-generating function* $M_X(t)$ of the random variable X is the expected value of the function e^{tx} where t is an auxiliary variable. Thus

$$M_X(t) = \langle e^{tX} \rangle. \tag{4.3.1}$$

When X is a continuous variable with probability density $p(x)$,

$$M_X(t) = \int_{-\infty}^{\infty} e^{tx} p(x) \, dx. \tag{4.3.2}$$

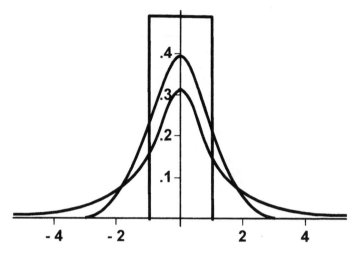

Figure 4.4. Probability densities of the uniform $U(0, 1)$, normal $N(0, 1)$, and Cauchy $C(0, 1)$ random variables.

Now, we can write the moments

$$\langle X^n \rangle = \int_{-\infty}^{\infty} dx p(x) x^n \tag{4.3.3}$$

as

$$\begin{aligned} \langle X^n \rangle &= \lim_{t \to 0} \int_{-\infty}^{\infty} dx p(x) \left(\frac{d}{dt}\right)^n (e^{tx}) \\ &= \lim_{t \to 0} \left(\frac{d}{dt}\right)^n \int_{-\infty}^{\infty} dx p(x) e^{tx} \\ &= \lim_{t \to 0} \left(\frac{d}{dt}\right)^n M_X(t). \end{aligned} \tag{4.3.4}$$

Thus, the moment $\langle X^n \rangle$ is the limit as $t \to 0$ of the nth derivative of $M_X(t)$ with respect to the auxiliary variable t. Taking derivatives is easier than doing integrations—hence the convenience.

For example, the moment-generating function of the uniform variable $U(m, a)$ is

$$\begin{aligned} M_U(t) &= \frac{1}{2a} \int_{m-a}^{m+a} e^{tx} \, dx \\ &= \frac{e^{t(m+a)} - e^{t(m-a)}}{2at}, \end{aligned} \tag{4.3.5}$$

and that of a normal variable $N(m, a^2)$ is

$$M_N(t) = \frac{1}{\sqrt{2\pi a^2}} \int_{-\infty}^{\infty} dx \exp\left[tx - \frac{(x-m)^2}{2a^2}\right]. \qquad (4.3.6)$$

By completing the square in the argument of the exponential, the latter reduces to

$$M_N(t) = \frac{\exp\left[mt + \frac{t^2 a^2}{2}\right]}{\sqrt{2\pi a^2}} \int_{-\infty}^{\infty} dx \exp\left[\frac{-(x-m-ta^2)^2}{2a^2}\right]. \qquad (4.3.7)$$

Given the substitution $u = x - m - ta^2$ (4.3.7) becomes

$$\begin{aligned} M_N(t) &= \frac{\exp\left[mt + \frac{t^2 a^2}{2}\right]}{\sqrt{2\pi a^2}} \int_{-\infty}^{\infty} du\, e^{-u^2/2a^2} \\ &= e^{mt + a^2 t^2/2}, \end{aligned} \qquad (4.3.8)$$

from which we can easily deduce expressions for the moments of a normal (see Problem 4.2, *Moments of a Normal*). Since only random variables with finite moments have a moment-generating function, the Cauchy variable $C(m, a)$ does not have one except in the special case when $a = 0$, in which case it collapses to the sure variable m. The moment-generating function of any sure variable m is $M_m(t) = \langle e^{mt} \rangle = e^{mt} \langle 1 \rangle = e^{mt}$.

When they exist, moment-generating functions completely define a random variable, or, alternatively, completely define its probability density. Showing that two random variables have the same moment-generating function is equivalent to showing that the two have identical probability densities, or that they, are *identically distributed*. Herein lies the moment-generating function's greatest theoretical utility. For instance, if two variables, X_1 and X_2, have the same moment-generating function, namely $e^{\mu t + \sigma^2 t^2/2}$, both are normal variables with mean μ and variance σ^2. We exploit this property of moment-generating functions in chapter 5. Recall, though, that two random variables can be identically distributed without being correlated.

Problems

4.1. Single-Slit Diffraction. According to the probability interpretation of light, formulated by Max Born in 1926, light intensity at a point is proportional to the probability that a photon exists at that point.

a. What is the probability density $p(x)$ that a single photon passes through a narrow slit and arrives at position x on a screen parallel to and at a distance d beyond the barrier? Each angle of forward propagation θ is the uniform random variable $U(0, \pi/2)$. See Figure 4.5. [Hint: Each differential range of realizations $(\theta + d\theta, \theta)$ maps into a differential range of realizations $(x + dx, x)$ in such a way that $p(\theta)d\theta = p(x)\, dx$, where the relationship between θ and x is clear from the geometry.]

b. The light intensity produced by diffraction through a single, narrow slit, as found in almost any introductory physics text, is proportional to

$$\frac{1}{r^2} \frac{\sin^2[(\pi a/\lambda)\sin\theta)]}{\sin^2\theta}$$

where r is the distance from the center of the slit to an arbitrary place on the screen, a is the slit width, and λ the light wavelength. Show that for slits so narrow that $\pi a/\lambda \ll 1$, the above light intensity is proportional to the photon probability density derived in part a.

4.2. Moments of a Normal.   Starting from the moment-generating function for $N(0, a^2)$, as provided in equation (4.3.8), show that $\langle N(0, \sigma^2)^n \rangle = 1 \cdot 3 \cdot 5 \ldots (n-1) \cdot \sigma^n$ for even n.

4.3. Exponential Random Variable.   Also according to Born's interpretation of light, the intensity of light exiting a slab of uniformly absorbing media is proportional to the probability that a photon will survive passage through the slab. If, as is reasonable to assume, the light absorbed $dI(x)$ in a differentially thin slab is proportional to its local intensity $I(x)$ and to the slab thickness dx, then $dI(x) = -\lambda I(x)dx$ and $I(x) \propto e^{-\lambda x}$. When normalized (on the semi-infinite line $x \geq 0$), the intensity of surviving photons becomes the photon probability density

$$p(x) = \lambda e^{-\lambda x} \qquad x \geq 0$$
$$ = 0 \qquad x < 0.$$

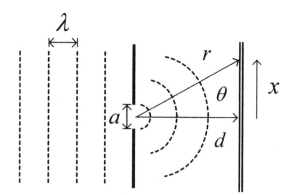

Figure 4.5. Single-slit diffraction.

The random variable so defined is called the *exponential random variable $E(\lambda)$*.

a. Show that mean$\{E(\lambda)\} = 1/\lambda$.
b. Find the moment-generating function $M_E(t)$ of $E(\lambda)$ for $t < \lambda$.
c. Use the moment-generating function to find var$\{E(\lambda)\}$.
d. Also, find $\langle E(\lambda)^n \rangle$ for arbitrary integer n.

4.4. Poisson Random Variable. The probability that n identical outcomes are realized in a very large set of statistically independent and identically distributed random variables when each outcome is extremely improbable is described by the Poisson probability distribution

$$P_n = \frac{e^{-\mu}\mu^n}{n!},$$

where $n = 0, 1, 2, 3, \ldots$ is the number of outcomes. For instance, the number of decays per second of a sample of the radioisotope U_{92}^{238} is a Poisson random variable, because the probability that any one nuclei will decay in a given second is very small and the number of nuclei within a macroscopic sample is very large. By definition, $\mu = \sum_{n=0}^{n=\infty} n P_n$, which one can demonstrate as

$$\sum_{n=0}^{\infty} n P_n = e^{-\mu} \sum_{n=0}^{\infty} \frac{\mu^{n+1}}{n!}$$

$$= \mu e^{-\mu} \sum_{n=0}^{\infty} \frac{\mu^n}{n!}$$

$$= \mu e^{-\mu} \left[1 + \mu + \frac{\mu^2}{2!} + \frac{\mu^3}{3!} + \cdots \right]$$

$$= \mu.$$

The last step follows from the Taylor series expansion,

$$e^{\mu} = 1 + \mu + \frac{\mu^2}{2!} + \frac{\mu^3}{3!} + \cdots.$$

a. Given that the average number of decays per second registered by a Geiger counter is 2, what is the probability that within a series of one-second rate measurements the number of decays per second will be 5?
b. Show that P_n is normalized—that is, show that

$$1 = \sum_{n=0}^{\infty} \frac{e^{-\mu}\mu^n}{n!}.$$

5

Normal Variable Theorems

5.1 Normal Linear Transform Theorem

Normal random variables have several properties that are especially valuable in applied statistics and random process theory. Here we formulate the normal linear transform theorem, the normal sum theorem, and the central limit theorem. In proving these theorems, we will exploit the properties of moment-generating functions.

According to the *normal linear transform theorem*, a linear function of a normal variable is another normal variable with appropriately modified mean and variance. Thus

$$\alpha + \beta N(m, a^2) = N(\alpha + \beta m, \beta^2 a^2). \tag{5.1.1}$$

If in (5.1.1) we set $m = 0$ and $a^2 = 1$, we have

$$\alpha + \beta N(0, 1) = N(\alpha, \beta^2). \tag{5.1.2}$$

Therefore, an arbitrary normal variable $N(\alpha, \beta^2)$ is a linear transform of a so-called *unit normal* $N(0, 1)$ with a mean of zero and a variance of one.

The proof of the normal linear transform theorem follows from identifying the moment-generating function of $\alpha + \beta N(m, a^2)$ with the moment-generating function of $N(\alpha + \beta m, \beta^2 a^2)$. For the former we have, by definition,

$$
\begin{aligned}
M_{\alpha+\beta N(m,a^2)}(t) &= \langle e^{t(\alpha+\beta N[m,a^2])} \rangle \\
&= e^{t\alpha} \langle e^{t\beta N(m,a^2)} \rangle \\
&= e^{t\alpha} M_{N(m,a^2)}(t\beta).
\end{aligned} \tag{5.1.3}
$$

From (4.3.8) we know that

$$M_{N(m,a^2)}(t) = e^{mt + \frac{a^2 t^2}{2}}, \tag{5.1.4}$$

and so

$$M_{N(m,a^2)}(t\beta) = e^{mt\beta + \frac{a^2 t^2 \beta^2}{2}}, \tag{5.1.5}$$

which when substituted into the right-hand side of (5.1.3) yields

$$M_{\alpha+\beta N(m,a^2)}(t) = e^{t(\alpha+\beta m)+\frac{\beta^2 a^2 t^2}{2}}. \tag{5.1.6}$$

The right-hand side of (5.1.6) is, by definition, the moment-generating function of $N(\alpha + \beta m, \beta^2 a^2)$. That is, (5.1.6) is equivalent to

$$M_{\alpha+\beta N(m,a^2)}(t) = M_{N(\alpha+\beta m, \beta^2 a^2)}(t). \tag{5.1.7}$$

Because the random variables $\alpha + \beta N(m, a^2)$ and $N(\alpha + \beta m, \beta^2 a^2)$ have the same moment-generating function, they are identically distributed, and, consequently, the normal linear transform theorem (5.1.1) is proved. See also Problem 5.1, *Uniform Linear Transform*.

5.2 Normal Sum Theorem

According to the *normal sum theorem*, two statistically independent normal variables sum to another normal variable. In particular,

$$N(m_1 + m_2, a_1^2 + a_2^2) = N_1(m_1, a_1^2) + N_2(m_2, a_2^2) \tag{5.2.1}$$

when $N_1(m_1, a_1^2)$ and $N_2(m_2, a_2^2)$ are statistically independent. The normal sum theorem is, of course, consistent with the already established fact (in sections 2.2 and 2.3) that the mean and variance of a sum of statistically independent random variables is the sum of the individual means and variances.

The proof of the normal sum theorem also follows from the properties of moment-generating functions. Suppose that $X_1 = N_1(m_1, a_1^2)$ and $X_2 = N_2(m_2, a_2^2)$. Then, according to (4.3.8), $M_{X_1}(t) = e^{m_1 t + \frac{a_1^2 t^2}{2}}$ and $M_{X_2}(t) = e^{m_2 t + \frac{a_2^2 t^2}{2}}$.

$$
\begin{aligned}
M_{X_1+X_2}(t) &= \langle e^{t(X_1+X_2)} \rangle \\
&= \langle e^{tX_1} \rangle \langle e^{tX_2} \rangle \\
&= M_{X_1}(t) M_{X_2}(t) \\
&= e^{t(m_1+m_2)+\frac{t^2(a_1^2+a_2^2)}{2}}
\end{aligned} \tag{5.2.2}
$$

where, in the second line, we have assumed that X_1 and X_2 are statistically independent. The right-hand side of (5.2.2) is now in the form of the moment-generating function of $N(m_1 + m_2, a_1^2 + a_2^2)$. Thus, the moment-generating function of $N(m_1, a_1^2) + N_2(m_2, a_2^2)$ is identical to the moment-generating function of $N(m_1 + m_2, a_1^2 + a_2^2)$, and the normal sum theorem for statistically independent addends has been proved.

Together, the normal linear transform and normal sum theorems establish that any linear function of statistically independent normal variables is another normal variable. Although uniform variables $U(m, a)$ and most random variables do not sum to their own kind, Cauchy variables $C(m, a)$ do so. See Problem 5.2, *Adding Uniform Variables*. The analog theorems for Cauchy variables are

$$\alpha + \beta C(m, a) = C(\alpha + \beta m, \beta a) \qquad (5.2.3)$$

and

$$C_1(m_1, a_1) + C_2(m_2, a_2) = C(m_1 + m_2, a_1 + a_2). \qquad (5.2.4)$$

The latter requires that $C_1(m_1, a_1)$ and $C_2(m_2, a_2)$ be statistically independent. Because $C(m, a)$ has infinite moments, we cannot prove (5.2.3) and (5.2.4) with moment-generating functions. The most direct proof of (5.2.3) and (5.2.4) exploits the so-called *random variable transform theorem* (Gillespie 1992).

5.3 Jointly Normal Variables

We can make an even more powerful statement: statistically dependent normals, if *jointly normal*, also sum to a normal. Two variables are jointly normal when they are each linear combinations of a single set of independent normals. To illustrate, consider the variables defined by

$$X_1 = a N_1(0, 1) \qquad (5.3.1)$$

and

$$X_2 = b N_1(0, 1) + c N_2(0, 1). \qquad (5.3.2)$$

Here a, b, and c are constants and $N_1(0, 1)$ and $N_2(0, 1)$ are, by specification, statistically independent unit normals. Here, as before, the different subscripts attached to $N(0, 1)$ denote statistical independence; identical subscripts would denote complete correlation. Thus, the variables X_1 and X_2 are, by definition, jointly normal. The property of joint normality covers a number of possible relationships. When $b \neq 0$ and $a \neq 0$, X_1 and X_2 are statistically dependent normal variables. When $c = 0$ and $a = b$, X_1 and X_2 are completely correlated, and, when $b = 0$, they are statistically independent. Yet, according to the normal sum (5.2.1) and linear transform (5.1.1) theorems,

$$
\begin{aligned}
X_1 + X_2 &= a N_1(0, 1) + b N_1(0, 1) + c N_2(0, 1) \\
&= (a + b) N_1(0, 1) + c N_2(0, 1) \\
&= N_1(0, (a + b)^2) + N_2(0, c^2) \\
&= N(0, (a + b)^2 + c^2). \qquad (5.3.3)
\end{aligned}
$$

Therefore, dependent but jointly distributed normals sum to a normal. See Problem 5.3, *Dependent Normals*.

Jointly normal variables arise naturally in the multivariate systems we discuss in chapters 8 and 9. There we exploit another of their special properties: two jointly normal variables,

$$Y_1 = a_0 + \sum_{i=1}^{m} a_i N_i(0, 1) \tag{5.3.4}$$

and

$$Y_2 = b_0 + \sum_{i=1}^{m} b_i N_i(0, 1), \tag{5.3.5}$$

are completely determined by only five moments: mean$\{Y_1\} = a_0$, var$\{Y_1\} = \sum_{i=1}^{m} a_i^2$, mean$\{Y_2\} = b_0$, var$\{Y_2\} = \sum_{i=1}^{m} b_i^2$, and cov$\{Y_1, Y_2\} = \sum_{i=1}^{m} a_i b_i$, even when the number m of independent unit normals $N_i(0, 1)$ out of which Y_1 and Y_2 are composed is larger than five. Thus, variations among the a_i and b_i which preserve these five quantities do not change Y_1 and Y_2.

The proof of this statement is beyond the scope of this book, but may be found in Springer (1979). Here we simply note that the probability density of two jointly normal variables Y_1 and Y_2 is

$$p(y_1 \& y_2) = \frac{1}{2\pi\sigma_1\sigma_2\sqrt{1-\rho^2}} e^{\frac{-1}{(1-\rho^2)}\left[\frac{(y_1-\mu_1)^2}{2\sigma_1^2} + \frac{(y_2-\mu_2)^2}{2\sigma_2^2} - \rho\frac{(y_1-\mu_1)(y_2-\mu_2)}{\sigma_1\sigma_2}\right]}. \tag{5.3.6}$$

For convenience, we have adopted the notation $\mu_1 = \text{mean}\{Y_1\}$, $\mu_2 = \text{mean}\{Y_2\}$, $\sigma_1^2 = \text{var}\{Y_1\}$, $\sigma_2^2 = \text{var}\{Y_2\}$, and $\rho = \text{cor}\{Y_1, Y_2\}$. Note that in (5.3.6) $p(y_1 \& y_2)$ has the expected property that when Y_1 and Y_2 are statistically independent, $\rho = 0$ and the joint probability density $p(y_1 \& y_2)$ factors into a product of two normal densities.

5.4 Central Limit Theorem

Can anything be said about the sum of random variables when the nature of the individual addends is not known? Amazingly, under certain conditions, the answer is yes. If the random variables $X_1, X_2, \ldots X_m$ are statistically independent, their means and variances finite, and their number m large, the sum

$$S_m = X_1 + X_2 + \cdots + X_m \tag{5.4.1}$$

is approximately normal with mean

$$\mu_m = \sum_{i=1}^{m} \text{mean}\{X_i\} \tag{5.4.2}$$

and variance

$$\sigma_m^2 = \sum_{i=1}^{m} \text{var}\{X_i\}. \tag{5.4.3}$$

Approximating the sum S_m with $N(\mu_m, \sigma_m^2)$ works well because the approximation is based on the *central limit theorem*, according to which

$$\lim_{m \to \infty} \frac{S_m - \mu_m}{\sigma_m} = N(0, 1) \qquad (5.4.4)$$

when the X_i composing the sum S_m are statistically independent and have finite means and variances. The central limit theorem is so called because it plays a central role in the statistical sciences.

Repeated addition turns statistically independent non-normal variables with finite means and variances into normal variables. Note, however, that the central limit theorem makes no claim about how quickly normality is approached as more terms are added to the sum S_m. One suspects that the closer to normal the addends X_i are, the more quickly S_m approaches normality. After all, normality is achieved with only two addends if the two are individually normal. Alternatively, if the addends are sufficiently non-normal—for example, if the addends are Cauchy variables $C(m, a)$—the central limit theorem doesn't apply and normality is never achieved.

We will prove the central limit theorem for the special case of identically distributed random addends $X_i (i = 1, 2, \ldots, m)$ for which moment-generating functions exist and so for which all moments $\langle X_i^n \rangle$ $(n = 1, 2, \ldots)$ are finite. Then it will follow that $\mu_0 = \langle X_i \rangle$ and $\sigma_0^2 = \langle X_i^2 \rangle - \langle X_i \rangle^2$ for all i. Consequently, $\mu_m = m\mu_0$ and $\sigma_m^2 = m\sigma_0^2$. As a first step, we form the random variable

$$Z_m = \frac{(S_m - \mu_m)}{\sigma_m}$$

$$= \frac{(S_m - m\mu_0)}{\sqrt{m\sigma_0^2}}. \qquad (5.4.5)$$

Given (5.4.1), we find that Z_m can be expressed as

$$Z_m = \sum_{i=1}^{m} \frac{(X_i - \mu_0)}{\sqrt{m\sigma_0^2}}$$

$$= \frac{1}{\sqrt{m}} \sum_{i=1}^{m} Y_i \qquad (5.4.6)$$

where the auxiliary variables

$$Y_i = \frac{(X_i - \mu_0)}{\sigma_0}, \qquad (5.4.7)$$

by design, have mean$\{Y_i\} = 0$ and var$\{Y_i\} = 1$.

The central limit theorem claims that Z_m approaches the unit normal $N(0, 1)$ as m becomes indefinitely large. Our strategy is to prove that the moment-generating function of Z_m approaches the moment-generating function of the unit normal $N(0, 1)$ as m becomes indefinitely large. From (5.4.6) and the definition of a moment-generating function (4.3.1), we find that

$$M_{Z_m}(t) = M_{\frac{1}{\sqrt{m}} \sum Y_i}(t)$$

$$= \left\langle e^{\frac{t}{\sqrt{m}} \sum_{i=1}^{m} Y_i} \right\rangle. \tag{5.4.8}$$

Because the X_i and, consequently, the Y_i are statistically independent,

$$M_{Z_m}(t) = \left\langle e^{\frac{tY_1}{\sqrt{m}}} \right\rangle \left\langle e^{\frac{tY_2}{\sqrt{m}}} \right\rangle \cdots \left\langle e^{\frac{tY_m}{\sqrt{m}}} \right\rangle. \tag{5.4.9}$$

Because the Y_i are identically distributed,

$$M_{Z_m}(t) = \left[\left\langle e^{\frac{tY_1}{\sqrt{m}}} \right\rangle \right]^m. \tag{5.4.10}$$

Expanding the exponential $e^{tY_1/\sqrt{m}}$ inside (5.4.10) in a Taylor series, we find that

$$M_{Z_m}(t) = \left[\left\langle 1 + \frac{tY_1}{\sqrt{m}} + \frac{t^2 Y_1^2}{2!m} + \frac{t^3 Y_1^3}{3!m^{3/2}} + \cdots \right\rangle \right]^m$$

$$= \left[1 + \frac{t^2}{2!m} + \frac{t^3 \langle Y_1^3 \rangle}{3!m^{3/2}} + \cdots \right]^m \tag{5.4.11}$$

since $\langle Y_1 \rangle = 0$ and $\langle Y_1^2 \rangle = 1$. Because all the moments $\langle Y_1^p \rangle$ are assumed finite, only the first two terms of the "multinomial" expansion (5.4.11) survive the $m \to \infty$ limit. Thus,

$$\lim_{m \to \infty} M_{Z_m}(t) = \lim_{m \to \infty} \left[1 + \frac{t^2}{2m} \right]^m \tag{5.4.12}$$

or, finally,

$$\lim_{m \to \infty} M_{Z_m}(t) = e^{t^2/2}. \tag{5.4.13}$$

The last step follows from the basic properties of the exponential function (Courant and Robbins 1941). Since $e^{t^2/2}$ is the moment-generating function of $N(0, 1)$, (5.4.13) proves the central limit theorem for identically distributed independent addends.

Many variables found in nature and conceived in physical models are sums of a large number of statistically independent variables, and thus are normal-like random variables. In chapter 6, we appeal to the central limit theorem in formulating the fundamental dynamical equations that govern random processes. The normal linear transform and normal sum theorems help us solve these dynamical equations.

Problems

5.1. Uniform Linear Transform. Prove $U(\alpha, \beta) = \alpha + \beta U(0, 1)$ by showing that $M_{U(\alpha,\beta)}(t) = M_{\alpha+\beta U(0,1)}(t)$.

5.2. Adding Uniform Variables. Prove that the sum $U_1(m_1, a_1) + U_2(m_2, a_2)$ of two statistically independent uniform variables $U_1(m_1, a_1)$ and $U_2(m_2, a_2)$ is not itself a uniform random variable by showing that the moment-generating function of $U_1(m_1, a_1) + U_2(m_2, a_2)$ is not in the form of a moment-generating function of a uniform random variable.

5.3. Dependent Normals. Given that $X_1 = aN_1(0, 1)$ and $X_2 = bN_1(0, 1) + cN_2(0, 1)$ where a, b, and c are constants and $N_1(0, 1)$ and $N_2(0, 1)$ are statistically independent unit normal variables, find

a. $\text{cov}\{X_1, X_2\}$,
b. $\text{var}\{X_1 + X_2\}$, and
c. $\text{var}\{X_1\} + \text{var}\{X_2\}$.
d. Show that $\text{var}\{X_1 + X_2\} \neq \text{var}\{X_1\} + \text{var}\{X_2\}$.

6

Einstein's Brownian Motion

6.1 Sure Processes

In large part, the goal of physics is to discover the time evolution of variables that describe important parts of the universe. By hypothesis, these variables are random variables. For instance, chapter 3 describes a model of the random position of a Brownian particle, but that model employs neither continuous random variables nor their continuous evolution in time. Chapters 4 and 5 have prepared us to work with continuously distributed random variables. Here we also investigate the consequences of assuming continuity in time. In preparation for that task, we first review important properties of continuous sure processes. Some of these properties carry over into random processes and some do not.

Consider the charge $q(t)$ on a capacitor of capacitance C shorted through a resistor of resistance R as illustrated in Figure 6.1.

Kirchoff's law,

$$i(t)R + \frac{q(t)}{C} = 0, \tag{6.1.1}$$

governs the process. Given that the current in the circuit $i(t)$ and the charge on the capacitor $q(t)$ are related by $i(t) = \frac{dq(t)}{dt}$, (6.1.1) becomes

$$dq(t) + \frac{q(t)}{RC}dt = 0, \tag{6.1.2}$$

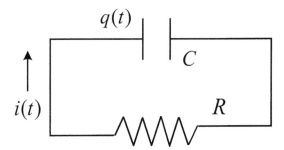

Figure 6.1. Charge $q(t)$ on a capacitor shorted through a resistor. The current $i(t)$ is $dq(t)/dt$.

which can also be written as

$$q(t + dt) - q(t) = -\frac{q(t)}{RC}dt, \tag{6.1.3}$$

where we have replaced the differential $dq(t)$ with its equivalent $q(t+dt)-q(t)$. The very form of (6.1.3) expresses continuity, smoothness, memorylessness, and determinism. Actually, two kinds of continuity are built into this dynamical equation. On the one hand, since time t is arbitrary and the increment dt can be made arbitrarily small, the process is *time-domain continuous*. On the other hand, since

$$\lim_{dt \to 0} q(t + dt) = q(t), \tag{6.1.4}$$

the process is *process-variable continuous*. The process is also *smooth* because the limit

$$\lim_{dt \to 0} \frac{q(t + dt) - q(t)}{dt} \tag{6.1.5}$$

exists. Here we deliberately treat the differential dt as if it is a small but finite quantity. Smoothness requires process-variable continuity, and process-variable continuity, in turn, requires time-domain continuity. However, a continuous process need not be smooth.

The process $q(t)$ is also a *memoryless* one, or, more commonly, a *Markov* process, because the value of $q(t)$ at any one instant, say at $t = t_1 + dt$, is determined by its value at $t = t_1$ through a dynamical equation, in this case (6.1.3) with t_1 replacing t. Alternatively stated, $q(t_1)$ alone predicts $q(t_1 + dt)$; no previous values $q(t_0)$ where $t_0 < t_1$ are needed. Most well-known processes in physics are Markov processes. Magnetic systems and others having long-term memory or *hysteresis* are exceptions. The Russian mathematician A. A. Markov (1856–1922) even used memoryless processes to model the occurrence of short words in the prose of the great Russian poet Pushkin.

Finally, the process $q(t)$ is *sure*, or *deterministic*, because equation (6.1.3) returns a unique value of $q(t + dt)$ for each $q(t)$.

Many of the familiar processes of classical physics belong to the class of time-domain and process-variable continuous, smooth, and Markov sure processes. In the next section we investigate a particular random process that is continuous (in both senses) and Markov but neither smooth nor sure. Such *continuous, Markov, random processes* incrementally, but powerfully, generalize the well-behaved, sure processes of classical physics they most closely resemble.

Although we don't explore them here, other kinds of random processes are both possible and useful (Gillespie 1992). In so-called *discrete time processes*, the time-domain on which the random variable is defined is a countable set of discrete times $\{t_0, t_1, \ldots\}$ such as might characterize different rounds of a game of chance or generations of a population. Numerical simulations of continuous processes are, necessarily, discrete time processes. In *jump processes*, the range of values the process variable can assume is countably discrete. For instance, the number of molecules $N(t)$ within a given permeable volume is defined on a continuous time interval, but $N(t)$ must equal some integer.

6.2 Wiener Process

A process variable $X(t)$ is defined by its associated probability density $p(x, t)$. Its two arguments, x and t, refer to the two different ways $X(t)$ can vary: in time t and in value x at each time. More specifically, $X(t)$ and $X(t + dt)$ are different random variables which, when representing different parts of a Markov process, are related by a dynamical equation of form

$$X(t + dt) - X(t) = F[X(t), dt]. \tag{6.2.1}$$

The *Markov propagator function* $F[X(t), dt]$ is itself a random variable and a function of a random variable. The propagator probabilistically determines $X(t + dt)$ from $X(t)$ via (6.2.1)—that is, in so far as one random variable can determine another. We assume time-domain and process variable continuity, so that $F[X(t), dt] \to 0$ as $dt \to 0$, but we do not require smoothness. The form (6.2.1) generalizes the sure processes of classical physics.

The *Wiener process*, defined by the Markov propagator

$$F[X(t), dt] = \sqrt{\delta^2 \, dt} N_t^{t+dt}(0, 1), \tag{6.2.2}$$

where δ^2 is a process-characterizing parameter, is the simplest of all continuous Markov processes. Its corresponding dynamical equation,

$$X(t + dt) - X(t) = \sqrt{\delta^2 \, dt} N_t^{t+dt}(0, 1), \tag{6.2.3}$$

is the basic unit out of which more complicated random processes are composed. Here $N_t^{t+dt}(0, 1)$ denotes a unit normal (with mean 0 and variance 1) associated explicitly with the time interval $(t, t + dt)$. Operationally, equation (6.2.3) means that when the Wiener process variable $X(t)$ realizes the sure value $x(t)$ at time t, $X(t + dt)$ is a normally distributed random variable with mean $x(t)$ and variance $\delta^2 \, dt$, or $X(t + dt) = N(x(t), \delta^2 dt)$. Alternatively, the realization $x(t + dt)$ is the sum of the sure variable $x(t)$ and the product of $\sqrt{\delta^2 \, dt}$ and a realization of the unit normal $N(0, 1)$.

The factor $\sqrt{dt}$ in the dynamical equation (6.2.3) seems odd. Are $\sqrt{dt}$ and dt allowed in the same differential equation? If one's standard is the ordinary calculus of sure processes, certainly not. Terms proportional to $\sqrt{dt}$ are indefinitely larger than terms proportional to dt as $dt \to 0$. However, here $\sqrt{dt}$ is multiplied by the unit normal $N_t^{t+dt}(0, 1)$, which in different subintervals assumes different positive and negative values. The net effect of adding them together is to reduce the magnitude of $\sqrt{dt}N_t^{t+dt}(0, 1)$ to that of dt. But note that as $dt \to 0$,

$$X(t + dt) \to X(t) \qquad (6.2.4)$$

and

$$\frac{dx}{dt} = \sqrt{\frac{\delta^2}{dt}}\, N_t^{t+dt}(0,1) \to \infty. \qquad (6.2.5)$$

Thus, the Wiener process is, on its domain, everywhere continuous but nowhere smooth. This special property makes the Wiener process dynamical equation (6.2.3) a different kind of mathematical object—a *stochastic differential equation*.

Time-domain continuity restricts possible interpretations of the Wiener process dynamical equation (6.2.3) and, as we shall see, encourages us to adopt the sub- and superscripts placed on the unit normal symbol $N_t^{t+dt}(0, 1)$. Because of time-domain continuity, $X(t + dt/2)$ exists and we can formally divide the process-variable increment $X(t + dt) - X(t)$ into the sum of two subincrements so that

$$X(t+dt) - X(t) = [X(t+dt) - X(t+dt/2)] + [X(t+dt/2) - X(t)]. \qquad (6.2.6)$$

While condition (6.2.6) seems trivial, it has a surprising consequence. Substituting the Wiener process propagator into both sides of (6.2.6) yields

$$\sqrt{\delta^2\, dt}N_t^{t+dt}(0, 1) = \sqrt{\delta^2(dt/2)}N_{t+\frac{dt}{2}}^{t+dt}(0, 1)$$

$$+ \sqrt{\delta^2(dt/2)}N_t^{t+\frac{dt}{2}}(0, 1), \qquad (6.2.7)$$

a condition that has been called *self-consistency* (Gillespie 1996). Self-consistency obtains only when the unit normals $N_{t+\frac{dt}{2}}^{t+dt}(0, 1)$ and $N_t^{t+\frac{dt}{2}}(0, 1)$ are statistically independent. As a rule, normals associated with temporally disjunct time intervals must be statistically independent in order that self-consistency and, ultimately, time-domain continuity be observed. In a phrase, $N_t^{t+dt}(0, 1)$ is *temporally uncorrelated*. If the time intervals are identical, the associated normals are completely correlated; if overlapping, statistically dependent; and if disjunct, statistically independent. For this reason—to remind us of their degree of mutual independence or dependence—we place sub- and superscripts on the unit normal symbol $N_t^{t+dt}(0, 1)$.

The Cauchy variable $C_t^{t+dt}(0, 1)$ also reduces terms of magnitude $\sqrt{dt}$ down to dt and satisfies self-consistency. In fact, the *Cauchy process* defined by

$$X(t + dt) - X(t) = \sqrt{\delta^2 dt}\, C_t^{t+dt}(0, 1) \qquad (6.2.8)$$

in many ways mimics the Wiener process (6.2.3). The relatively long tails on the probability density associated with $C(0, 1)$ make longer excursions in $X(t)$ possible at the expense of many shorter ones. Normal and Cauchy variables are only two members of a class of *Lévy variables*, named after the French mathematician Paul Lévy (1886–1971), who studied their properties. Each Lévy variable preserves its nature under linear transformation and addition. For this reason, each can also be made the basis of a Lévy process with properties similar to the Wiener and Cauchy processes. Applications range from the seemingly random flight of the albatross to particle motion in turbulent media (Klafter et al., 1996 and 1999).

Yet normal processes are the only continuous Markov processes that produce random variables with finite variances, and finite variances are often required for physical interpretation. For instance, the variance of a random velocity $V(t)$ is related to the mean kinetic energy, and the latter must be finite. The central limit theorem also favors normal processes. One can imagine a process that, on the smallest time scales, is composed of non-normal but statistically independent increments with finite means and variances. A large number of these subscale increments sum, via the central limit theorem, to a propagator that is approximately continuous and normal on time scales of interest.

6.3 Brownian Motion Revisited

The Wiener process is the perfect mathematical vehicle for describing continuous Brownian motion. Suppose, as in chapter 3, the Brownian particle moves in one dimension along the x-axis. The net effect of many molecular impacts is to displace the particle an amount

$$X(t + dt) - X(t) = \sqrt{\delta^2 \, dt}\, N_t^{t+dt}(0, 1) \qquad (6.3.1)$$

in the interval $(t, t + dt)$. These displacements are indifferently positive and negative, with size regulated by the parameter δ^2.

How does the net displacement of the Brownian particle evolve with time? We now have the tools to integrate the stochastic differential equation (6.3.1) and answer this question. When $t = 0$, (6.3.1) becomes

$$X(dt) = X(0) + \sqrt{\delta^2 \, dt}\, N_0^{dt}(0, 1), \qquad (6.3.2)$$

and when $t = dt$,

$$X(2dt) = X(dt) + \sqrt{\delta^2 dt} N_{dt}^{2dt}(0, 1). \tag{6.3.3}$$

Dropping the former into the right-hand side of the latter produces

$$X(2dt) = X(0) + \sqrt{\delta^2 \, dt} N_0^{dt}(0, 1) + \sqrt{\delta^2 \, dt} N_{dt}^{2dt}(0, 1). \tag{6.3.4}$$

Because $N_0^{dt}(0, 1)$ and $N_{dt}^{2dt}(0, 1)$ apply to disjunct time intervals, they are statistically independent, and the two terms on the far right of (6.3.4) sum, via the normal sum and linear transform theorems, to

$$X(2dt) = X(0) + N_0^{2dt}(0, \delta^2 2dt). \tag{6.3.5}$$

Repeating this substitution and addition indefinitely produces

$$X(t) = X(0) + N_0^t(0, \delta^2 t). \tag{6.3.6}$$

Thus $X(t) - X(0)$ is normally distributed with a vanishing mean and, as in chapter 3, a variance that grows linearly in time t. But note that here a single parameter δ^2 has replaced the quotient $\Delta x^2 / \Delta t$ of two independently specified parameters Δx^2 and Δt.

6.4 Monte Carlo Simulation

A Brownian particle, initially at the origin, occupies the position

$$X(t) = N_0^t(0, 1)\sqrt{\delta^2 t} \tag{6.4.1}$$

at time t. But how does X evolve in time between 0 and t? One could evaluate (6.4.1) at a series of intermediate times $0, t/n, 2t/n, 3t/n, \ldots t$ where $n > 1$ and so produce a sequence of position variables

$$X(0) = 0,$$

$$X\left(\frac{t}{n}\right) = N_0^{t/n}(0, 1)\sqrt{\delta^2 \frac{t}{n}},$$

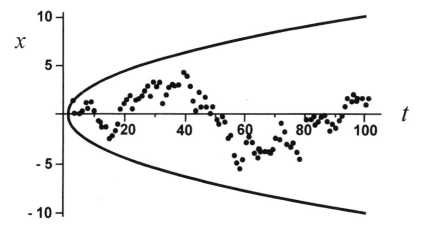

Figure 6.2. The dots represent a realization of the Wiener process $X(t) = N_0^t(0, t)$, determined every time steop of size $\Delta t = 1$ for 100 steps by solving the update equation (6.4.3) recursively. The solid line is a one-standard deviation envelope $\pm\sqrt{t}$.

$$X\left(\frac{2t}{n}\right) = N_0^{2t/n}(0, 1)\sqrt{\delta^2\frac{2t}{n}},$$
$$\dots$$
$$X(t) = N_0^t(0, 1)\sqrt{\delta^2 t}. \tag{6.4.2}$$

But a special problem arises if one wants to produce realizations of these variables: the unit normals $N_0^{t/n}(0, 1)$, $N_0^{2t/n}(0, 1)$, ... $N_0^t(0, 1)$ are mutually dependent, and the process $X(t)$ is autocorrelated. See Problem 6.1, *Autocorrelated Process*. Self-consistency can be used to link the correlated variables in (6.4.2), but usually one accounts for autocorrelation with a different method: by numerically advancing the particle position with an *update equation*,

$$X(t + \Delta t) = x(t) + N_t^{t+\Delta t}(0, 1)\sqrt{\delta^2 \Delta t}, \tag{6.4.3}$$

derived by replacing t in the exact solution (6.3.6) with $t + \Delta t$ and applying the initial condition $X(t) = x(t)$. A *Monte Carlo simulation* is simply a sequence of such updates with the realization of the updated position $x(t + \Delta t)$ at the end of each time step used as the initial position $x(t)$ at the beginning of the next. Figure 6.2 was produced in this way. The 100 plotted points mark sample positions along the particle's trajectory. Equally valid, if finer-scaled, *sample paths* could be obtained with smaller time steps Δt. But recall that $X(t)$ is not a smooth process and its time derivative does not exist. For this reason it would be misleading to connect the points in figure 6.2 with a smooth curve.

6.5 Diffusion Equation

The probability density

$$p(x, t) = \frac{1}{\sqrt{2\pi \delta^2 t}} e^{-\frac{x^2}{2\delta^2 t}} \tag{6.5.1}$$

defines the random variable $N_0^t[0, \delta^2 t]$. Figure 6.3, displaying $p(x, t)$ versus x at times $t = 1/4$, 1, and 4, illustrates the possibilities inherent in the time evolution of a Wiener process more completely, if more abstractly, than the sample path of figure 6.2. Inspecting the partial derivatives

$$\frac{\partial}{\partial t} p(x, t) = -\frac{p(x, t)}{2t} \left[1 - \frac{x^2}{\delta^2 t} \right], \tag{6.5.2}$$

$$\frac{\partial}{\partial x} p(x, t) = -\frac{x}{\delta^2 t} p(x, t), \tag{6.5.3}$$

and

$$\frac{\partial^2}{\partial x^2} p(x, t) = -\frac{p(x, t)}{\delta^2 t} \left[1 - \frac{x^2}{\delta^2 t} \right], \tag{6.5.4}$$

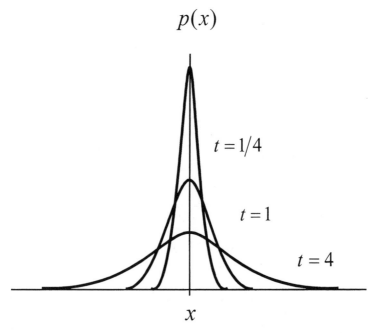

$$p(x)$$

$$t = 1/4$$

$$t = 1$$

$$t = 4$$

$$x$$

Figure 6.3. The probability density $p(x, t) = (2\pi \delta^2 t)^{-1/2} \exp\{-x^2/2\delta^2 t\}$ at times $t = 1/4$, 1, and 4.

we find that $p(x, t)$ solves the classical *diffusion equation*

$$\frac{\partial p(x, t)}{\partial t} = \frac{\delta^2}{2} \frac{\partial^2 p(x, t)}{\partial x^2}. \tag{6.5.5}$$

Equation (6.5.5) is mathematically equivalent to the stochastic dynamical equation (6.3.1). The latter equation governs the random variable $X(t)$, while the former governs its probability density $p(x, t)$. Deducing the diffusion equation (6.5.5) from its solution (6.5.1) reverses the usual order in modeling and problem solving. A more physically motivated derivation of (6.5.5) often starts with the observation, called *Fick's law*, that a gradient in the probability density $\partial p / \partial x$ drives a probability density flux J so that

$$J = -D \frac{\partial p}{\partial x}. \tag{6.5.6}$$

where the proportionality constant D is called the *diffusion constant*. Fick's law, like $F = ma$ and $V = IR$, both defines a quantity (diffusion constant, mass, or resistance) and states a relation between variables. The diffusion constant is positive definite, that is, $D \geq 0$, because a gradient always drives an oppositely directed flux in an effort to diminish the gradient. Combining Fick's law and the one-dimensional conservation or *continuity equation*

$$\frac{\partial p}{\partial t} + \frac{\partial J}{\partial x} = 0 \tag{6.5.7}$$

yields the diffusion equation (6.5.5) with D replacing $\delta^2/2$.

In his famous 1905 paper on Brownian motion, Albert Einstein (1879–1955) constructed the diffusion equation in yet another way—directly from the continuity and Markov properties of Brownian motion. Our approach, in section 6.3, to the mathematically equivalent result $X(t) - X(0) = N_0^t(0, 2Dt)$ has been via the algebra of random variables. We use the phrase *Einstein's Brownian motion* to denote both these configuration-space descriptions (involving only position x or X) of Brownian motion. In chapters 7 and 8, we will explore their relationship to Newton's Second Law and possible velocity-space descriptions (involving velocity v or V as well as position).

Problems

6.1. Autocorrelated Process. Let $X(t)$ and $X(t')$ be the instantaneous random position of a Brownian particle at times for which $t' \leq t$.

a. Find $\text{cov}\{X(t), X(t')\}$.
b. Find $\text{cor}\{X(t), X(t')\}$.
c. Evaluate $\text{cor}\{X(t), X(t')\}$ in the limits $t'/t \to 0$ and $t'/t \to 1$.

(Hint: Refer to the solution [6.3.6] and to self-consistency [6.2.7]. Also compare with Problem 3.4, *Autocorrelation*.)

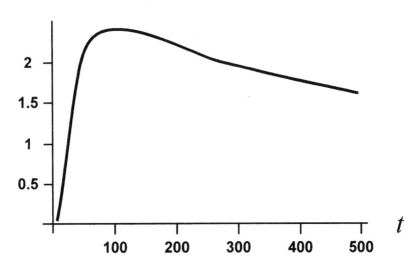

Figure 6.4. Local particle density $N_0 p(x, t)$ versus time at $x = x_1 > 0$, given that all the particles are initialized at $x = 0$. Here $\delta^2 = 1$, $x_1 = 10$, and $N_0 = 100$.

6.2. Concentration Pulse. Suppose that N_0 particles of dye are released at time $t = 0$ in the center (at $x = 0$) of a fluid contained within an essentially one-dimensional pipe, and the dye is allowed to diffuse in both directions along the pipe. The diffusion constant $D = \delta^2/2$. At position $X(t)$ and time t the density of dye particles is the product $N_0 p(x, t)$, where $p(x, t)$ is the probability density of a single dye particle with initialization $X(0) = 0$. An observer at position $x = x_1 \neq 0$ sees the concentration of dye increase to a maximum value and then decay away. See figure 6.4. At what time does the concentration peak pass the observer?

6.3. Brownian Motion with Drift. Consider the dynamical equation $X(t + dt) - X(t) = \alpha\, dt + \sqrt{\delta^2\, dt}\, N_t^{t+dt}(0, 1)$, describing Brownian motion superimposed on a steady drift of rate α.

a. Given the initial condition $X(0) = 0$, solve this equation using the method in section 6.3.

b. Find the associated probability density $p(x, t)$.

c. Show that the full width of $p(x, t)$ at half its maximum value increases in time as $2\sqrt{2\delta^2 t \ln 2}$.

Because the center of $p(x, t)$ evolves as αt and its full width at half maximum evolves more slowly as $2\sqrt{2\delta^2 t \ln 2}$, it is possible to separate different species

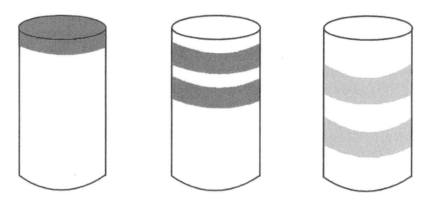

Figure 6.5. Sedimentation: layers of Brownian particles drifting downward and diffusing in a viscous fluid. Time increases to the right.

of Brownian particles with different drift rates α. Figure 6.5 illustrates this separation in the context of *sedimentation*. In similar fashion, *electrophoresis* uses an electric field to separate charged Brownian particles (Berg 1993).

6.4. Brownian Motion in a Plane.   Use solutions $X(t) = N_{0,x}^t(0, 1)\sqrt{\delta^2 t}$ and $Y(t) = N_{0,y}^t(0, 1)\sqrt{\delta^2 t}$ and the method in section 6.4 to generate and plot a Brownian particle sample path in the x-y plane. Assume the unit normals $N_{0,x}^t(0, 1)$ and $N_{0,y}^t(0, 1)$ (and thus displacements in the two directions) are statistically independent.

7

Ornstein-Uhlenbeck Processes

7.1 Langevin Equation

Newton's second law identifies the net force $F(t)$ per unit particle mass M with the rate at which the particle changes its velocity $V(t)$. This velocity, in turn, describes the rate at which the particle changes its position $X(t)$. These familiar relations,

$$\frac{dV(t)}{dt} = \frac{F(t)}{M} \tag{7.1.1}$$

and

$$\frac{dX(t)}{dt} = V(t), \tag{7.1.2}$$

are no less true when $V(t)$ and $X(t)$ are random variables than otherwise. In differential form, we have

$$V(t + dt) - V(t) = \left[\frac{F(t)}{M}\right] dt \tag{7.1.3}$$

and

$$X(t + dt) - X(t) = V(t) \, dt. \tag{7.1.4}$$

Albert Einstein and his French contemporary Paul Langevin (1872–1946) introduced randomness into these equations in different ways.

Einstein's actual analysis resulted in the diffusion equation (6.5.5), but we now know that in his procedure he essentially ignored Newton's second law (7.1.3) and replaced $V(t) \, dt$ on the right-hand side of (7.1.4) with $\sqrt{\delta^2 \, dt} \; N_t^{t+dt}(0, 1)$ (Einstein, 1905). This replacement turns (7.1.4) into

$$X(t + dt) - X(t) = \sqrt{\delta^2 \, dt} N_t^{t+dt}(0, 1), \tag{7.1.5}$$

and thus turns $X(t)$ into a Wiener process with parameter δ^2.

Attacking the same problem a few years later, Paul Langevin modeled the specific impulse $[F(t)/M] \, dt$ in Newton's second law (7.1.3) as a viscous drag

$-\gamma V(t)\,dt$ plus random fluctuations $\sqrt{\beta^2\,dt}Z_t$. According to Langevin, the random variable Z_t had mean zero, variance one, was "indifferently positive and negative," and was uncorrelated with position $X(t)$. If one specifies that $Z_t = N_t^{t+dt}(0, 1)$, equation (7.1.3) becomes the *Langevin equation*,

$$V(t + dt) - V(t) = -\gamma V(t)\,dt + \sqrt{\beta^2\,dt}N_t^{t+dt}(0, 1). \qquad (7.1.6)$$

The Langevin equation is said to govern an *Ornstein-Uhlenbeck* or *O-U process*, after L. S. Ornstein and G. E. Uhlenbeck, who formalized the properties of this continuous Markov process (Uhlenbeck and Ornstein 1930). The O-U process $V(t)$ and its time integral $X(t)$ together describe Langevin's Brownian motion.

Langevin's main insight was that viscous drag and velocity fluctuations are complementary effects of a single, subscale phenomenon: numerous, frequent collisions between fluid molecules and the Brownian particle. The same collisions in the same interval contribute to the fluctuation term $\sqrt{\beta^2\,dt}N_t^{t+dt}(0, 1)$ and to the viscous drag term $-\gamma V(t)\,dt$. The former, no less than the latter, represents the effect of many collisions. It may be for this reason that Langevin referred to the fluctuating term in (7.1.6) as the "complementary force." An English translation of Langevin's landmark paper appears in Appendix A. In this chapter we solve the Langevin equation, quantify the link between drag (or dissipation) γ and fluctuation β^2 constants, and model electrical noise with an O-U process before returning, in chapter 8, to complete the description of Langevin's Brownian motion.

7.2 Solving the Langevin Equation

We could directly integrate the Langevin equation (7.1.6) to find an expression for $V(t)$ just as we integrated, in section 6.3, the stochastic differential equation describing Einstein's Brownian motion to find $X(t)$. We would do this by recursively evaluating the stochastic differential equation (7.1.6) at different times and summing the parts. However, the sums are difficult—in part because the addends are correlated. Instead, we adopt a simpler and more powerful method for solving stochastic differential equations. This new method is based on the following logic. Since each variable in the sequence of random variables $V(dt)$, $V(2\,dt)$, ..., $V(t)$ is a linear combination of the independent normal variables $N_0^{dt}(0, 1)$, $N_{dt}^{2dt}(0, 1)$, ..., $N_{t-dt}^{t}(0, 1)$ and linear combinations of statistically independent normals are themselves normal, then $V(t)$ is itself normal, that is,

$$V(t) = N_0^t(\text{mean}\{V(t)\}, \text{var}\{V(t)\}). \qquad (7.2.1)$$

So our problem reduces to finding expressions for the sure functions mean$\{V(t)\}$ and var$\{V(t)\}$ and substituting these into the form (7.2.1).

Taking the expected value of both sides of the Langevin equation (7.1.6) produces an ordinary differential equation whose solution is mean$\{V(t)\}$. Thus

$$\langle V(t+dt) - V(t)\rangle = \langle -\gamma V(t)\,dt + \sqrt{\beta^2\,dt}N_t^{t+dt}(0, 1)\rangle, \qquad (7.2.2)$$

and

$$\langle V(t+dt)\rangle - \langle V(t)\rangle = -\gamma\langle V(t)\rangle\,dt + \sqrt{\beta^2\,dt}\langle N_t^{t+dt}(0, 1)\rangle$$

$$= -\gamma\langle V(t)\rangle\,dt, \qquad (7.2.3)$$

or, equivalently,

$$\frac{d\langle V(t)\rangle}{dt} = -\gamma\langle V(t)\rangle, \qquad (7.2.4)$$

where we have exploited the linearity of the expected value operator $\langle\rangle$ and the fact that $\langle N_t^{t+dt}(0, 1)\rangle = 0$. Solving the ordinary differential equation (7.2.4), we find that

$$\text{mean}\{V(t)\} = v_0 e^{-\gamma t} \qquad (7.2.5)$$

given the initial condition $V(0) = v_0$.

The time evolution of var$\{V(t)\}$, or, equivalently, of $\langle V(t)^2\rangle - \langle V(t)\rangle^2$, also follows from the Langevin equation but less directly so. Since, from (7.2.5), we already know that $\langle V(t)\rangle^2 = v_0^2 e^{-2\gamma t}$, we only need find $\langle V(t)^2\rangle$. By definition,

$$d[V(t)^2] = [V(t+dt)]^2 - [V(t)]^2. \qquad (7.2.6)$$

The Langevin equation (7.1.6) provides an expression for $V(t+dt)$ that, when substituted into (7.2.6), yields

$$d[V(t)^2] = [V(t)(1 - \gamma dt) + \sqrt{\beta^2\,dt}N_t^{t+dt}(0, 1)]^2 - [V(t)]^2$$

$$= V(t)^2(1 - \gamma\,dt)^2 + 2V(t)(1 - \gamma dt)\sqrt{\beta^2\,dt}N_t^{t+dt}(0, 1)$$

$$\quad + \beta^2\,dt[N_t^{t+dt}(0, 1)]^2 - V(t)^2$$

$$= -2V(t)^2\gamma dt + 2V(t)\sqrt{\beta^2\,dt}N_t^{t+dt}(0, 1)$$

$$\quad + \beta^2\,dt[N_t^{t+dt}(0, 1)]^2, \qquad (7.2.7)$$

where we have dropped terms of order dt^2 and $dt^{3/2}$ because they are ignorably small compared to dt. Taking the expected value of (7.2.7) produces

$$d\langle V(t)^2 \rangle = -2\langle V(t)^2 \rangle \gamma dt + 2\langle V(t)N_t^{t+dt}(0, 1)\rangle \sqrt{\beta^2 \, dt}$$
$$+ \langle [N_t^{t+dt}(0, 1)]^2 \rangle \beta^2 \, dt$$
$$= -2\langle V(t)^2 \rangle \gamma dt + 2\langle V(t)N_t^{t+dt}(0, 1)\rangle \sqrt{\beta^2 \, dt}$$
$$+ \beta^2 \, dt. \tag{7.2.8}$$

Recall that $V(t)$ is a linear combination of $N_o^{dt}(0, 1)$, $N_{dt}^{2dt}(0, 1)$, ... and $N_{t-dt}^t(t)$ but not of $N_t^{t+dt}(t)$. Thus, $V(t)$ and $N_t^{t+dt}(t)$ are statistically independent, and

$$\langle V(t)N_t^{t+dt}(0, 1)\rangle = \langle V(t)\rangle\langle N_t^{t+dt}(0, 1)\rangle$$
$$= 0. \tag{7.2.9}$$

Then (7.2.8) becomes

$$d\langle V(t)^2 \rangle = -2\langle V(t)^2 \rangle \gamma dt + \beta^2 dt \tag{7.2.10}$$

or

$$\frac{d}{dt}\langle V(t)^2 \rangle = -2\gamma \langle V(t)^2 \rangle + \beta^2. \tag{7.2.11}$$

Solving (7.2.11) subject to the initial condition $V(0) = v_0$ yields

$$\langle V(t)^2 \rangle = v_0^2 e^{-2\gamma t} + \left(\frac{\beta^2}{2\gamma}\right)(1 - e^{-2\gamma t}). \tag{7.2.12}$$

With practice one learns to streamline these manipulations. Combining $\langle V(t)\rangle^2$ and $\langle V(t)^2 \rangle$ from (7.2.5) and (7.2.12), we find

$$\text{var}\{V(t)\} = \langle V(t)^2 \rangle - \langle V(t)\rangle^2$$
$$= \left(\frac{\beta^2}{2\gamma}\right)(1 - e^{-2\gamma t}), \tag{7.2.13}$$

as is consistent with the expected initial condition $\text{var}\{V(0)\} = 0$. Substituting expressions for $\text{mean}\{V(t)\}$ from (7.2.5) and $\text{var}\{V(t)\}$ from (7.2.13) into the normal variable form (7.2.1) yields the desired O-U process solution

$$V(t) = N_0^t \left(v_0 e^{-\gamma t}, \left(\frac{\beta^2}{2\gamma}\right)(1 - e^{-2\gamma t})\right). \tag{7.2.14}$$

The corresponding probability density is

$$p(v,t) = \frac{\exp\left[\dfrac{-(v - v_0 e^{-\gamma t})^2}{2(\beta^2/2\gamma)(1 - e^{-2\gamma t})}\right]}{\sqrt{2\pi\left(\dfrac{\beta^2}{2\gamma}\right)(1 - e^{-2\gamma t})}}. \tag{7.2.15}$$

These describe not only the velocity of a Brownian particle but also any process governed by a competition between linear damping and constant magnitude random fluctuations. See, for instance, the discussion in section 7.5 on thermally generated electrical noise and that in section 10.1 on molecular effusion.

Here we note two useful generalizations. First, suppose the initial condition was left an unspecified random variable. Then,

$$V(t) = V(0) + N_0^t\left(\langle V(0)\rangle(e^{-\gamma t} - 1), \left(\frac{\beta^2}{2\gamma}\right)(1 - e^{-\gamma t})\right) \tag{7.2.16}$$

solves the Langevin equation. Second, suppose the Langevin equation included a drift v_d to which $V(t)$, apart from fluctuations, relaxed in the long time limit, so that (7.1.6) is replaced by

$$V(t + dt) - V(t) = -\gamma[V(t) - v_d]\,dt + \sqrt{\beta^2\,dt}\,N_t^{t+dt}(0, 1). \tag{7.2.17}$$

Such drift v_d might be the terminal velocity caused by gravity, by an electric field, or by any other constant force in the presence of dissipation. The solution of (7.2.17) is

$$V(t) = N_0^t\left(v_d + e^{-\gamma t}(v_0 - v_d), \left(\frac{\beta^2}{2\gamma}\right)(1 - e^{-2\gamma t})\right), \tag{7.2.18}$$

given $V(0) = v_0$. See Problem 7.1, *Terminal Speed* for an application of the O-U process with drift.

7.3 Simulating the O-U Process

Because the random variable $V(t)$ is autocorrelated, numerically simulating the O-U process is not simply a matter of substituting a time t into the desired form of the solution, choosing a realization of the unit normal, and calculating the result. The best way to account numerically for potentially strong autocor-

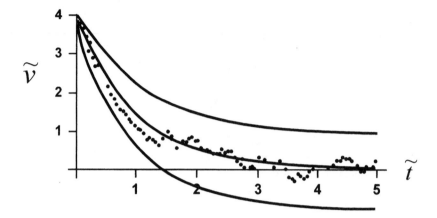

Figure 7.1. Points on a sample path of the normalized O-U process defined by (7.3.2) with initial value $\tilde{v}_o = 4$ and drift $\tilde{v}_d = 0$. Solid curves show mean$\{\tilde{V}(\tilde{t})\}$ and mean$\{\tilde{V}(\tilde{t})\} \pm \sqrt{\text{var}\{\tilde{V}(\tilde{t})\}}$.

relation is, as in section 6.4, to write the general solution (7.2.18) in updated form

$$V(t + \Delta t) = v(t)e^{-\gamma \Delta t} + v_d(1 - e^{-\gamma \Delta t})$$

$$+ \sqrt{\left(\frac{\beta^2}{2\gamma}\right)(1 - e^{-2\gamma \Delta t})} N_t^{t+\Delta t}(0, 1), \qquad (7.3.1)$$

in which the initial condition $v(t)$ is a particular realization of the process variable $V(t)$ determined in the previous interval.

Before proceeding, we recast (7.3.1) in terms of the following dimensionless variables and parameters: $\tilde{t} = \gamma t$, $\Delta\tilde{t} = \gamma \Delta t$, $\tilde{V}(t) = V(t)/\sqrt{\beta^2/2\gamma}$, $\tilde{v}_d = v_d/\sqrt{\beta^2/2\gamma}$, and $\tilde{v}(t) = v(t)/\sqrt{\beta^2/2\gamma}$. Then (7.3.1) becomes

$$\tilde{V}(\tilde{t} + \Delta\tilde{t}) = \tilde{v}(\tilde{t})e^{-\Delta\tilde{t}} + \tilde{v}_{\tilde{d}}(1 - e^{-\Delta\tilde{t}}) + \sqrt{(1 - e^{-2\Delta\tilde{t}})} N_{\tilde{t}}^{\tilde{t}+\Delta\tilde{t}}(0, 1), \quad (7.3.2)$$

from which we have formally eliminated γ and β^2. Thus one sample path generated by recursively solving (7.3.2) works for all values of γ and β^2. One simply reinterprets the meaning of $\tilde{t}$, $\Delta\tilde{t}$, $\tilde{V}(\tilde{t})$, $\tilde{v}_d$, and $\tilde{v}(\tilde{t})$. Figure 7.1 (with $\tilde{v}_0 = 4$ and $\tilde{v}_d = 0$) and figure 7.2 (with $\tilde{v}_0 = 0$ and $\tilde{v}_d = 3$) display such sample paths.

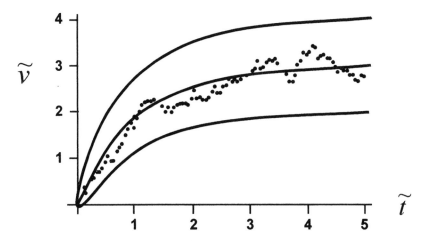

Figure 7.2. Points on a sample path of the normalized O-U process defined by (7.3.2) with initial value $\tilde{v}_0 = 0$ and drift $\tilde{v}_d = 3$. Solid curves show mean$\{\tilde{V}(\tilde{t})\}$ and mean$\{\tilde{V}(\tilde{t})\} \pm \sqrt{\text{var}\{\tilde{v}(\tilde{t})\}}$

7.4 Fluctuation-Dissipation Theorem

Competition between linear damping and random fluctuations defines the O-U process. In the long time limit $\gamma t \to \infty$, a balance is achieved between the two, and the average kinetic energy of a Brownian particle of mass M in the frame in which the drift vanishes becomes, according to (7.2.13),

$$\frac{M \, \text{var}\{V(\infty)\}}{2} = \frac{M\beta^2}{4\gamma}. \tag{7.4.1}$$

But in the same limit ($\gamma t \to \infty$), the Brownian particle also approaches thermal equilibrium with the surrounding fluid. According to the *equipartition theorem*, the equilibrium energy associated with fluctuations in each degree of freedom is $kT/2$, where T is the fluid temperature. Thus, it must be that

$$\frac{M \, \text{var}\{V(\infty)\}}{2} = \frac{kT}{2}. \tag{7.4.2}$$

Therefore, the O-U process is consistent with thermal equilibrium only if

$$\frac{M\beta^2}{4\gamma} = \frac{kT}{2}, \tag{7.4.3}$$

that is, only if $\beta^2/2\gamma = kT/M$. Equation (7.4.3) is one version of the *fluctuation-dissipation theorem*, so named because it relates fluctuation β^2 and dissipation γ parameters.

Fluctuation-dissipation applies to any O-U process when fluctuations are caused by interaction with an environment that is itself in thermal equilibrium. In practice, fluctuation-dissipation helps fix the parameters β^2 and γ. When modeling a Brownian particle, one usually choses a dissipation rate γ and then solves (7.4.3) for β^2. For instance, when the Brownian particle is immersed in a viscous liquid, as assumed by Einstein and Langevin, Stokes's law

$$\gamma = \frac{6\pi \eta r}{M} \tag{7.4.4}$$

applies. Here η is the liquid viscosity and r the particle radius. Consequently, $\beta^2 = 12\pi \eta r kT/M^2$. On the other hand, if the fluid is composed of gas molecules with mass m_0 and density n_0 colliding with the Brownian particle at a rate $n_0\sigma v_{th}$, determined by the molecule-particle cross section σ and the gas thermal velocity $v_{th} = \sqrt{kT/m_0}$, then

$$\gamma = \frac{m_0 n_0 \sigma v_{th}}{M}. \tag{7.4.5}$$

In this case, fluctuation-dissipation yields $\beta^2 = 2m_0 n_0 \sigma v_{th} kT/M^2$.

7.5 Johnson Noise

Consider how the electrostatic energy stored on a charged capacitor dissipates when the capacitor is shorted through a resistor, as illustrated in figure 7.3. As charge carriers flow through the circuit, they collide with, and transfer energy to, the atoms of the resisting material. Eventually, the resistor shares this dissipated energy with the environment. However, since the resistor is not at absolute zero, its atoms contain thermal energy, which makes them vibrate around their equilibrium positions and randomly transfer energy to the charge carriers. Where there is dissipation, there is fluctuation. In the language of macro-scale physics: the resistor simultaneously Joule heats *and* delivers random voltage pulses to the circuit. The random pulses, first observed by J. B. Johnson in 1928, are called *Johnson noise*.

Johnson noise is easily modeled with an O-U process. Applying Kirchoff's law to an RC circuit with a fluctuating voltage source yields

$$IR + \frac{Q}{C} + (Johnson \cdot noise) = 0 \tag{7.5.1}$$

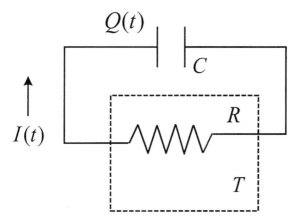

Figure 7.3. Capacitance C shorted through a resistance R at temperature T. The current $I(t)$ is $dQ(t)/dt$.

or, given that $I = dQ/dt$,

$$dQ = -\frac{Q}{RC}\,dt - \frac{(Johnson \cdot noise)}{R}\,dt. \qquad (7.5.2)$$

If the Johnson noise voltage fluctuations are described by a Wiener process with parameter β^2, (7.5.2) becomes

$$dQ = -\frac{Q}{RC}\,dt + \sqrt{\beta^2\,dt}\,N_t^{t+dt}(0, 1). \qquad (7.5.3)$$

This a Langevin equation with relaxation rate $\gamma = 1/RC$ and fluctuation parameter β^2, so we may take its solution,

$$Q(t) = N_0^t\left(q_0 e^{-\gamma t}, \left(\frac{\beta^2}{2\gamma}\right)(1 - e^{-2\gamma t})\right), \qquad (7.5.4)$$

directly from (7.2.14). The longtime, steady-state variance of the charge fluctuations, $\beta^2/2\gamma$, must be consistent with thermal equilibrium at temperature T. According to the equipartition theorem, the mean fluctuating electrostatic energy stored at equilibrium in the capacitor is given by $\mathrm{var}\{Q(\infty)\}/2C = kT/2$. Combining these requirements, we have, as before, the fluctuation-dissipation theorem, $B^2/4\gamma C = kT/2$, which, on using $\gamma = 1/RC$, yields $\beta^2 = 2kT/R$.

Therefore, in terms of circuit parameters R and C, the charge on the capacitor is given by

$$Q(t) = N_0^t(q_0 e^{-t/RC}, kTC(1 - e^{-2t/RC})). \qquad (7.5.5)$$

For an application of Johnson noise in a slightly different context, see Problem 7.2 *RL Circuit*.

Problems

7.1. Terminal Speed.   One way to determine the viscous drag parameter γ is to apply a steady force F to a particle of mass M and measure its mean terminal speed v_d.

   a. Express γ in terms of F, M, and v_d.
   b. Use the fluctuation-dissipation theorem to express the fluctuation parameter β^2 in terms of kT, F, M, and v_d where T is the fluid temperature.

7.2. RL Circuit.   Use energy equipartition to show that, in a circuit composed of an inductance L shorted through a resistance R at equilibrium temperature T, equilibrium current fluctuations have a mean $\langle I(\infty) \rangle = 0$ and variance $\langle I(\infty)^2 = kT/L$.

8

Langevin's Brownian Motion

8.1 Integrating the O-U Process

The O-U process $V(t)$ and its integral $X(t)$ together describe Langevin's Brownian motion. Given the velocity $V(t)$ of a Brownian particle, how do we find its position $X(t)$? We might try substituting

$$V(t) = N_0^t \left(v_0 e^{-\gamma t} + v_d(1 - e^{-\gamma t}), \left(\frac{\beta^2}{2\gamma} \right)(1 - e^{-2\gamma t}) \right), \qquad (8.1.1)$$

into

$$X(t + dt) - X(t) = V(t)dt \qquad (8.1.2)$$

and solving iteratively. Such procedure generates a series of expressions for $X(dt)$, $X(2dt)$, ..., $X(t)$, each one of which is a linear combination of correlated unit normals $N_0^{dt}(0, 1)$, $N_0^{2dt}(0, 1)$, ..., $N_0^{t-dt}(0, 1)$. Each of these is, in turn, a linear combination of statistically independent unit normals $N_0^{dt}(0, 1)$, $N_{dt}^{2dt}(0, 1)$, ..., $N_{t-2dt}^{t-dt}(0, 1)$. While, in principle, it might be possible to unpack these linear combinations, it is, as before, easier to exploit the general result based on the normal sum theorem that

$$X(t) = N_0^t(\text{mean}\{X(t)\}, \text{var}\{X(t)\}). \qquad (8.1.3)$$

Recall from section 5.3 that any pair of correlated normals, say $X(t)$ and $V(t)$, is completely determined by their means, variances, and a covariance. The O-U process (8.1.1) provides us with expressions for mean$\{V(t)\}$ and var$\{V(t)\}$. Thus, our task reduces to finding and solving the ordinary differential equations governing mean$\{X(t)\}$, var$\{X(t)\}$, and cov$\{X(t), V(t)\}$.

For convenience, in the following we replace $X(t + dt) - X(t)$ with dX and $X(t + dt)$ with $X + dX$, and we, likewise, replace $V(t + dt) - V(t)$ with dV

and $V(t + dt)$ with $V + dV$. With these substitutions the Langevin equation with drift V_d becomes

$$dV = -\gamma(V - v_d)dt + \sqrt{\beta^2 dt}\, N_t^{t+dt}(0, 1) \qquad (8.1.4)$$

and (8.1.2) becomes

$$dX = Vdt. \qquad (8.1.5)$$

Combining the expected value of $dX = Vdt$ and of V (from [8.1.1]) generates the differential equation

$$\frac{d\langle X \rangle}{dt} = \langle V \rangle$$
$$= v_0 e^{-\gamma t} + v_d(1 - e^{-\gamma t}), \qquad (8.1.6)$$

whose solution is

$$\text{mean}\{X(t)\} = x_0 + \frac{v_0}{\gamma}(1 - e^{-\gamma t}) + \frac{v_d}{\gamma}(\gamma t + e^{-\gamma t} - 1) \qquad (8.1.7)$$

assuming initial conditions $V(0) = v_0$ and $X(0) = x_0$. We already see a difference between Einstein's simple result, $\text{mean}\{X(t)\} = x_0 + v_d t$, and Langevin's more complicated one (8.1.7).

The equation governing $\text{var}\{X(t)\}$ also follows from $dX = Vdt$ and the expression (8.1.1) for $V(t)$. By definition, $dX(t)^2 = X(t + dt)^2 - X(t)^2$, which, in our streamlined notation, becomes

$$dX^2 = (X + dX)^2 - X^2$$
$$= 2XdX + (dX)^2$$
$$= 2XVdt + (Vdt)^2. \qquad (8.1.8)$$

Taking the expected value of (8.1.8), dividing by dt, and taking the limit $dt \to 0$ produces

$$\frac{d\langle X^2 \rangle}{dt} = 2\langle XV \rangle. \qquad (8.1.9)$$

Consequently,

$$\frac{d\,\text{var}\{X\}}{dt} = \frac{d}{dt}[\langle X^2 \rangle - \langle X \rangle^2]$$

$$= \frac{d\langle X^2 \rangle}{dt} - 2\langle X \rangle \frac{d\langle X \rangle}{dt}$$

$$= 2\langle XV \rangle - 2\langle X \rangle \langle V \rangle$$

$$= 2\,\mathrm{cov}\{X, V\}. \tag{8.1.10}$$

Thus, var$\{X\}$ couples to the as yet unknown function cov$\{X, V\}$. In deriving the equation governing cov$\{X, V\}$, we retain terms through order dX, dV, $(dV)^2$, and dt and drop terms of order $(dX)^2$, $dXdV$, $(dt)^{3/2}$, and smaller because these vanish after dividing by dt and taking the limit $dt \to 0$. Consequently,

$$
\begin{aligned}
d\,\mathrm{cov}\{X, V\} &= d[\langle XV \rangle - \langle X \rangle \langle V \rangle] \\
&= \langle X\,dV \rangle + \langle V\,dX \rangle - \langle X \rangle d\langle V \rangle - \langle V \rangle d\langle X \rangle \\
&= -\gamma \langle XV \rangle dt + \langle X N_t^{t+dt}(0, 1) \rangle \sqrt{\beta^2 dt} + \langle V^2 \rangle dt \\
&\quad + \gamma \langle X \rangle \langle V \rangle dt - \langle V \rangle^2 dt, \tag{8.1.11}
\end{aligned}
$$

where we have used $dX = Vdt$, $d\langle X \rangle = \langle V \rangle dt$, $d\langle V \rangle = -\gamma \langle V \rangle dt$, and Langevin's equation (8.1.4) for dV. Equation (8.1.11) simplifies to

$$
\begin{aligned}
d\,\mathrm{cov}\{X, V\} &= -\gamma\,\mathrm{cov}\{X, V\}dt + \mathrm{var}\{V\}dt \\
&\quad + \langle X N_t^{t+dt}(0, 1) \rangle \sqrt{\beta^2 dt}. \tag{8.1.12}
\end{aligned}
$$

If not identically zero, the term $\langle X N_t^{t+dt}(0, 1) \rangle \sqrt{\beta^2 dt}$ in (8.1.12) would dominate over the others because $\sqrt{dt}$ is very much larger than dt. However, because $X(t)$ and $N_t^{t+dt}(0, 1)$ are statistically independent, $\langle X N_t^{t+dt}(0, 1) \rangle = \langle X \rangle \langle N_t^{t+dt}(0, 1) \rangle = 0$, and (8.1.12) reduces to the ordinary differential equation

$$\frac{d}{dt}\,\mathrm{cov}\{X, V\} = -\gamma\,\mathrm{cov}\{X, V\} + \mathrm{var}\{V\}. \tag{8.1.13}$$

Multiplying through by an integrating factor $e^{\gamma t}$ turns (8.1.13) into

$$\frac{d}{dt}[e^{\gamma t}\,\mathrm{cov}\{X, V\}] = e^{\gamma t}\,\mathrm{var}\{V\}, \tag{8.1.14}$$

which, given that var$\{V\} = (\beta^2/2\gamma)(1 - e^{-2\gamma t})$ from (8.1.1) and initial conditions $V(0) = v_0$ and $X(0) = x_0$, integrates to

$$\text{cov}\{X, V\} = \frac{\beta^2}{2\gamma^2}(1 - 2e^{-\gamma t} + e^{-2\gamma t}). \qquad (8.1.15)$$

Substituting this result into the differential equation (8.1.10) for var$\{X\}$ produces

$$\frac{d}{dt} \text{var}\{X\} = \frac{\beta^2}{\gamma^2}(1 - 2e^{-\gamma t} + e^{-2\gamma t}), \qquad (8.1.16)$$

which is immediately integrated to yield

$$\text{var}\{X\} = \frac{\beta^2}{\gamma^2}\left[t - \frac{2}{\gamma}(1 - e^{-\gamma t}) + \frac{1}{2\gamma}(1 - e^{-2\gamma t}) \right]. \qquad (8.1.17)$$

Collecting the results (8.1.7) and (8.1.17), we have

$$X(t) = N_0^t\left(x_0 + \frac{v_0}{\gamma}(1 - e^{-\gamma t}) + \frac{v_d}{\gamma}(\gamma t + e^{-\gamma t} - 1), \right.$$

$$\left. \frac{\beta^2}{\gamma^3}\left[t\gamma - 2(1 - e^{-\gamma t}) + \frac{1}{2}(1 - e^{-2\gamma t}) \right] \right). \qquad (8.1.18)$$

This expression, together with (8.1.1) for $V(t)$ and (8.1.15) for cov$\{X(t), V(t)\}$, completely describes Langevin's Brownian motion with drift v_d. Note that only after many relaxation times (that is, when $\gamma t \gg 1$) does var$\{X(t)\}$ have the linear time dependence $\beta^2 t/\gamma^2$ characteristic of Einstein's Brownian motion.

8.2 Simulating Langevin's Brownian Motion

Our purpose in this section is to derive a simulation algorithm for an O-U process. This is equivalent to deriving an expression for the updated quantities $V(t + \Delta t)$ and $X(t + \Delta t)$ in terms of the initial values $V(t)$ and $X(t)$ and the O-U process parameters β^2 and γ. Since $V(t + \Delta t)$ and $X(t + \Delta t)$ are jointly distributed normals, they are correlated and, therefore, can be cast into the form

$$X(t + \Delta t) = a_0 + a_1 N_1(0, 1) + a_2 N_2(0, 1) \qquad (8.2.1)$$

and

$$V(t + \Delta t) = b_0 + b_1 N_1(0, 1). \qquad (8.2.2)$$

The unit normals $N_1(0, 1)$ and $N_2(0, 1)$ are, by design, statistically independent. Furthermore, the parameters $a_0, a_1, a_2, b_0,$ and b_1 must be chosen in order to give $X(t+\Delta t)$ and $V(t+\Delta t)$ the right values of mean$\{X(t+\Delta t)\}$, mean$\{V(t+\Delta t)\}$, var$\{X(t+\Delta t)\}$, var$\{V(t+\Delta t)\}$, and cov$\{X(t+\Delta t), V(t+\Delta t)\}$. Taking moments of (8.2.1) and (8.2.2), we find that

$$a_0 = \text{mean}\{X(t + \Delta t)\}, \qquad (8.2.3)$$

$$b_0 = \text{mean}\{V(t + \Delta t)\}, \qquad (8.2.4)$$

$$b_1^2 = \text{var}\{V(t + \Delta t)\}, \qquad (8.2.5)$$

$$a_1 = \frac{\text{cov}\{X(t + \Delta t), V(t + \Delta t)\}}{\sqrt{\text{var}\{V(t + \Delta t)\}}}, \qquad (8.2.6)$$

and

$$a_2^2 = \text{var}\{X(t + \Delta t)\} - \frac{(\text{cov}\{X(t + \Delta t), V(t + \Delta t)\})^2}{\text{var}\{V(t + \Delta t)\}}. \qquad (8.2.7)$$

See Problem 8.1 *Derivation*. The time dependences of mean$\{X(t + \Delta t)\}$, mean$\{V(t + \Delta t)\}$, var$\{X(t + \Delta t)\}$, var$\{V(t + \Delta t)\}$, and cov$\{X(t + \Delta t), V(t + \Delta t)\}$ follow directly from expressions (8.1.1), (8.1.15), and (8.1.18), already derived for mean$\{X(t)\}$, mean$\{V(t)\}$, var$\{X(t)\}$, var$\{V(t)\}$, and cov$\{X(t), V(t)\}$. In particular, we find that

$$\text{mean}\{V(t + \Delta t)\} = v(t)e^{-\gamma\Delta t} + v_d(1 - e^{-\gamma\Delta t}), \qquad (8.2.8)$$

$$\text{var}\{V(t + \Delta t)\} = \frac{\beta^2}{2\gamma}(1 - e^{-2\gamma\Delta t}), \qquad (8.2.9)$$

$$\text{mean}\{X(t + \Delta t)\} = x(t) + \frac{v(t)}{\gamma}(1 - e^{-\gamma\Delta t})$$

$$+ \frac{v_d}{\gamma}(\gamma\Delta t + e^{-\gamma\Delta t} - 1) \qquad (8.2.10)$$

$$\text{var}\{X(t + \Delta t)\} = \frac{\beta^2}{\gamma^3}(\gamma\Delta t) - \frac{2\beta^2}{\gamma^3}(1 - e^{-\gamma\Delta t})$$

$$+ \frac{\beta^2}{2\gamma^3}(1 - e^{-2\gamma\Delta t}), \qquad (8.2.11)$$

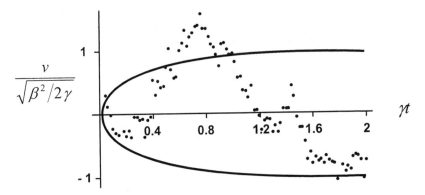

Figure 8.1. Points represent normalized sample velocities $v/\sqrt{\beta^2/2\gamma}$ versus normalized time γt of a Brownian particle from (8.2.2) with drift $v_d = 0$ and initial condition $v_0 = 0$. Solid curves represent mean$\{v/\sqrt{\beta^2/2\gamma}\}$ and mean$\{v/\sqrt{\beta^2/2\gamma}\} \pm$ std$\{v/\sqrt{\beta^2/2\gamma}\}$.

and

$$\text{cov}\{X(t + \Delta t), V(t + \Delta t)\} = \frac{\beta^2}{2\gamma^2}(1 - 2e^{-\gamma \Delta t} + e^{-2\gamma \Delta t}), \qquad (8.2.12)$$

which, when substituted into (8.2.3) through (8.2.7), provide the sought-for simulation algorithm for Langevin's Brownian motion in time steps of duration Δt.

Figures 8.1 and 8.2 display sample time evolutions $x(t)$ and $v(t)$ generated by solving (8.2.1) and (8.2.2) iteratively with $v_d = 0$ and initial conditions $x_0 = 0$ and $v_0 = 0$. As expected, when $v(t) > 0$, $x(t)$ increases in time, when $v(t) < 0$, $x(t)$ decreases in time. Also, as expected, $x(t)$ appears to evolve smoothly in time while $v(t)$ does not.

8.3 Smoluchowski Approximation

Langevin's Brownian motion reduces to Einstein's in the so-called *Smoluchowski limit*, that is, on time scales for which the independent variable $V(t)$ changes little and its integral $X(t)$ changes much. The Smoluchowski approximation effectively minimizes inertial effects and maximizes randomness. Formally, we access this regime by setting $dV = 0$ in the Langevin equation

$$dV = -\gamma V dt + \sqrt{\beta^2 dt}\, N_t^{t+dt}(0, 1) \qquad (8.3.1)$$

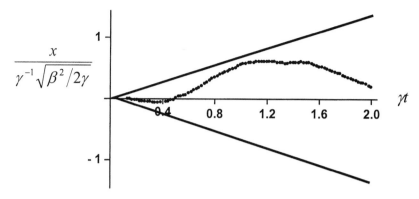

Figure 8.2. Points represent normalized sample positions $x/[\gamma^{-1}\sqrt{\beta^2/2\gamma}]$ versus normalized time γt of a Brownian particle from (8.2.1) and (8.2.2) with drift $v_d = 0$ and initial condition $x_0 = 0$. Solid curves represent $\text{mean}\{x/\gamma^{-1}\sqrt{\beta^2/2\gamma}\}$ and $\text{mean}\{x/\gamma^{-1}\sqrt{\beta^2/2\gamma}\} \pm \text{std}\{x/\gamma^{-1}\sqrt{\beta^2/2\gamma}\}$.

and using $dX = V dt$ to eliminate the variable V altogether. In this way (8.3.1) reduces to the Wiener process equation

$$dX = \sqrt{\frac{\beta^2}{\gamma^2}} dt\, N_t^{t+dt}(0, 1). \tag{8.3.2}$$

Its solution

$$X(t) = N_0^t\left(x_0, \frac{\beta^2 t}{\gamma^2}\right) \tag{8.3.3}$$

reproduces Einstein's Brownian motion, with β^2/γ^2 playing the role of the diffusion parameter δ^2. Apart from a constant offset in the mean position, solution (8.3.3) realizes the late-time, high-dissipation regime ($\gamma t \gg 1$) of Langevin's Brownian motion as described by (8.1.18) with $v_d = 0$.

8.4 Example: Brownian Projectile

A neutral molecule and a droplet of uncombusted gas thrust from a car tail pipe both obey the same stochastic dynamics—both are, in principle, Brownian particles evolving as an O-U process. Their vastly different masses account for quantitative rather than qualitative differences. In both cases the fluctuation-dissipation theorem fixes $\beta^2/2\gamma$ at kT/M. Then the dissipation rate γ and the thermal velocity $\sqrt{kT/M}$ alone adjust the degree to which, at any time t, the Brownian particle manifests either random or deterministic behavior.

Suppose, for instance, the Brownian particle is initialized as a projectile (with $x_0 = y_0 = z_0 = 0$, $v_{x0} = v_{y0} \neq 0$, and $v_{z0} = 0$) and moves under the influence of gravity ($\underline{g} = -g\hat{y}$) in the $x - y$ plane. The equations of motion are

$$dV_x = -\gamma V_x dt + \sqrt{\beta^2 dt}\, N_{t,x}^{t+dt}(0, 1), \qquad (8.4.1)$$

$$dV_y = -\gamma (V_y - v_d)dt + \sqrt{\beta^2 dt}\, N_{t,y}^{t+dt}(0, 1), \qquad (8.4.2)$$

$dX = V_x dt$, and $dY = V_y dt$. Here $v_d = -g/\gamma$ and the unit normals, $N_{t,x}^{t+dt}(0, 1)$ and $N_{t,y}^{t+dt}(0, 1)$, associated with fluctuations in different directions, are statistically independent. The solutions

$$X(t) = N_{0,x}^t \left(\frac{v_{x,0}}{\gamma}(1 - e^{-\gamma t}), \frac{\beta^2}{\gamma^3} \left[t\gamma - 2(1 - e^{-\gamma t}) + \frac{1}{2}(1 - e^{-2\gamma t}) \right] \right)$$
$$(8.4.3)$$

and

$$Y(t) = N_{0,y}^t \left(\frac{v_{y,0}}{\gamma}(1 - e^{-\gamma t}) - \frac{g}{\gamma^2}(\gamma t + e^{-\gamma t} - 1), \right.$$
$$\left. \frac{\beta^2}{\gamma^3} \left[t\gamma - 2(1 - e^{-\gamma t}) + \frac{1}{2}(1 - e^{-2\gamma t}) \right] \right) \qquad (8.4.4)$$

are taken from (8.1.18).

These configuration space coordinates reveal deterministic behavior at early times, that is, at times for which $\gamma t \ll 1$. In particular, through leading order in the assumed small quantity γt,

$$\text{mean}\{X(t)\} = v_{x,0}t \qquad (8.4.5)$$

and

$$\text{mean}\{Y(t)\} = v_{y,0}t - \frac{gt^2}{2}, \qquad (8.4.6)$$

which are familiar from introductory physics, and

$$\text{var}\{X(t)\} = \text{var}\{Y(t)\} = \frac{\beta^2 t^3}{3}. \qquad (8.4.7)$$

Thus, the early-time regime preserves the effect of initial conditions and reproduces familiar projectile motion, and the variance grows relatively slowly with

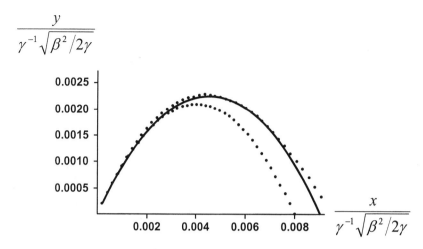

Figure 8.3. Early-time ($0 \leq \gamma t \leq 0.01$) sample trajectories (points) and mean trajectory (solid) for a Brownian particle under the influence of a normalized gravity of magnitude $(g\gamma^{-1}/\sqrt{\beta^2/2\gamma} = 220$. Initial conditions are $x_0 = y_0 = 0$ and $v_{x0} = v_{y0} = \sqrt{\beta^2/2\gamma}$.

time t. In contrast, at late times (when $\gamma t \gg 1$), the coordinates change more randomly. Through leading order terms in $1/\gamma t$,

$$\text{mean}\{X(t)\} = 0, \tag{8.4.8}$$

$$\text{mean}\{Y(t)\} = -\frac{gt}{\gamma}, \tag{8.4.9}$$

and

$$\text{var}\{X(t)\} = \text{var}\{Y(t)\} = \frac{\beta^2 t}{\gamma^2}. \tag{8.4.10}$$

Thus, at late times Brownian motion is superimposed on a downward constant drift.

Simulations of the processes (8.4.3) and (8.4.4) are displayed in figures 8.3 and 8.4 as trajectories in the x-y plane. Figure 8.3 shows a pair of largely deterministic and projectilelike trajectories. The trajectory of figure 8.4 passes through deterministic to random Brownian motion. In spite of the resemblance of figure 8.3 to familiar ballistic trajectories, the model producing them applies only at low speeds in viscous media—that is, only to Brownian projectiles. Typically, the drag force on baseballs and other macroscopic objects in air is not linear but quadratic in the speed.

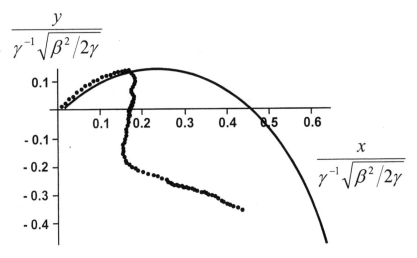

Figure 8.4. Early- through late-time $(0 \leq \gamma t \leq 1.0)$ sample trajectory (points) and mean trajectory (solid) for a Brownian particle under the influence of a normalized gravity of magnitude $(g\gamma^{-1}/\sqrt{\beta^2/2\gamma} = 3$. Initial conditions are $x_0 = y_0 = 0$ and $v_{x0} = v_{y0} = \sqrt{\beta^2/2\gamma}$.

The Langevin equation and its direct extensions are well suited for weaving deterministic and random effects together. In chapter 9 we investigate stochastic models of two other multivariate systems with familiar deterministic limits: the harmonic oscillator and the magnetized charged particle.

Problems

8.1. Derivation.   Derive (8.2.3) through (8.2.7) from (8.2.1) through (8.2.2).

8.2. X-V Correlation.   Find cor$\{X(t), V(t)\}$ for Langevin's Brownian motion in the late time, high-dissipation regime, that is, through leading order in the assumed large quantity γt.

8.3. Range Variation.   A Brownian particle starts at the origin and completes projectilelike motion in the X-Y plane under the influence of gravity while in its deterministic phase $(\gamma t \ll 1)$. Its time of flight t_f is the nonzero solution of mean$\{Y(t_f)\} = 0$.

   a. Express var$\{X(t_f)\}$ in terms of fluctuation parameter β^2, relaxation rate γ, acceleration of gravity g, and initial vertical velocity v_{y0}.

b. Given that $v_{x0} = v_{y0} = \sqrt{\beta^2/2\gamma}$, find an expression for the ratio of $std\{X(t_f)\}$ to the distance $\gamma^{-1}\sqrt{\beta^2/2\gamma}$ in terms of V_{yo}, γ, and g.

c. Numerically evaluate the dimensionless ratio $std\{X(t_f)\}/[\gamma^{-1}\sqrt{\beta^2/2\gamma}]$ with parameters used in producing figure 8.3. Is your result consistent with that of figure 8.3? Note: the sample paths of figure 8.3 suggest that $std\{X(t_f)\}/[\gamma^{-1}\sqrt{\beta^2/2\gamma}] \approx 10^{-3}$.

9

Other Physical Processes

9.1 Stochastic Damped Harmonic Oscillator

Imagine a massive object attached to a spring and submerged in a viscous fluid, as illustrated in figure 9.1. If set in motion, the object moves back and forth with an amplitude that slowly decays in time. The deterministic equations of motion governing this system,

$$dv = -\omega^2 x dt - \gamma v dt \qquad (9.1.1)$$

and

$$dx = v dt \qquad (9.1.2)$$

are those of a damped harmonic oscillator with oscillation frequency ω and decay rate γ. Yet the collisions causing the oscillations to decay also cause the oscillator to fluctuate randomly. The simplest self-consistent equations of motion describing a stochastically damped harmonic oscillator are

$$dV = -\omega^2 X dt - \gamma V dt + \sqrt{\beta^2 dt} N_t(0, 1) \qquad (9.1.3)$$

and

$$dX = V dt. \qquad (9.1.4)$$

The symbol $N_t(0, 1)$ in (9.1.3) slightly abbreviates previous notation $N_t^{t+dt}(0, 1)$ for the temporally uncorrelated unit normal associated with the time interval $(t, t + dt)$. The $\beta \to 0$ limit of (9.1.3) formally recovers the deterministic equation of motion (9.1.1), but, as before, the fluctuation-dissipation theorem requires that β and γ be related through $\beta^2/2\gamma = kT/M$, where M is the object mass and T the fluid temperature. Thus, only when the thermal speed $\sqrt{kT/M}$ is ignorably small compared to the oscillator speed is a deterministic description appropriate.

Solving (9.1.3) and (9.1.4) is a demanding but worthwhile task, first accomplished by Subrahmanyan Chandrasekhar (1910–95) in 1943. The harmonic oscillator, much more than the stationary particle or the free-falling projectile, is

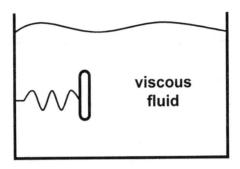

Figure 9.1. Viscously damped harmonic oscillator.

the ideal multipurpose tool of theoretical physics. When displaced a little from stable equilibrium, almost any object vibrates as a damped harmonic oscillator, and models of complex objects are often constructed out of weakly interacting oscillators. Furthermore, the harmonic oscillator has been a means of exploring the physics of new phenomena—nonlinearity, quantum mechanics, and, here, stochasticity.

The methods of chapter 8 suffice. Because equations (9.1.3) and (9.1.4) are linear stochastic differential equations, $X(t)$ and $V(t)$ are linear combinations of a set of uncorrelated normals $N_t(0, 1)$. Thus

$$X(t) = N(\text{mean}\{X(t)\}, \text{var}\{X(t)\}) \tag{9.1.5}$$

and

$$V(t) = N(\text{mean}\{V(t)\}, \text{var}\{V(t)\}) \tag{9.1.6}$$

where $\text{cov}\{X(t), V(t)\} \neq 0$. As before, our task reduces to finding and solving ordinary differential equations governing the time dependence of the means, $\text{mean}\{X\}$ and $\text{mean}\{V\}$, the variances, $\text{var}\{X\}$ and $\text{var}\{V\}$, and the covariance $\text{cov}\{X, V\}$. The equations for $\langle X \rangle$ and $\langle V \rangle$,

$$\frac{d\langle V \rangle}{dt} = -\omega^2 \langle X \rangle - \gamma \langle V \rangle \tag{9.1.7}$$

and

$$\frac{d\langle X \rangle}{dt} = \langle V \rangle, \tag{9.1.8}$$

are solved in many classical mechanics texts. These texts usually distinguish among *lightly damped* ($\gamma < 2\omega$), *critically damped* ($\gamma = 2\omega$), and *strongly damped* $(\gamma > 2\omega)$ behaviors. The solutions

$$\langle X(t) \rangle = e^{-\gamma t/2} \left[x_0 \cos(\omega' t) + \left(v_0 + \frac{\gamma x_0}{2} \right) \frac{\sin(\omega' t)}{\omega'} \right] \tag{9.1.9}$$

and

$$\langle V(t) \rangle = e^{-\gamma t/2} \left[v_0 \cos(\omega' t) - \left(x_0 \omega^2 + \frac{\gamma v_0}{2} \right) \frac{\sin(\omega' t)}{\omega'} \right] \tag{9.1.10}$$

apply in all three regimes but are written here to emphasize the lightly damped case. The reduced frequency $\omega' = \sqrt{\omega^2 - \gamma^2/4}$ is, of course, only real and positive definite when the oscillator is lightly damped.

The ordinary differential equation governing var$\{V(t)\}$ follows from the equation of motion (9.1.3) by way of

$$dV^2 = (V + dV)^2 - V^2$$

$$= 2V dV + (dV)^2, \tag{9.1.11}$$

$$d\langle V^2 \rangle = -2\gamma \langle V^2 \rangle dt - 2\omega^2 \langle VX \rangle dt + \beta^2 dt, \tag{9.1.12}$$

and

$$\frac{d\langle V^2 \rangle}{dt} = -2\gamma \langle V^2 \rangle - 2\omega^2 \langle VX \rangle + \beta^2. \tag{9.1.13}$$

Combining (9.1.13) and (9.1.7), we find the desired equation

$$\frac{d \, \text{var}\{V\}}{dt} = -2\gamma \, \text{var}\{V\} - 2\omega^2 \, \text{cov}\{X, V\} + \beta^2. \tag{9.1.14}$$

The equation governing the time dependence of var$\{X(t)\}$ follows, in like manner, from $dX = V dt$ via $dX^2 = 2X dX + (dX)^2$, $dX^2 = 2XV dt$, $d\langle X^2 \rangle = 2\langle XV \rangle dt$, and $d\langle X \rangle/dt = \langle V \rangle$. From these we find

$$\frac{d \, \text{var}\{X\}}{dt} = 2 \, \text{cov}\{X, V\}. \tag{9.1.15}$$

Also, from $d(XV) = (X + dX)(V + dV) - XV$ we have $d(XV) = X dV + V dX$ and, consequently, $d\langle XV \rangle = -\omega^2 \langle X^2 \rangle dt - \gamma \langle XV \rangle dt + \langle V^2 \rangle dt$, which, given

(9.1.7) and (9.1.8), becomes

$$\frac{d \, \text{cov}\{X, V\}}{dt} = -\gamma \, \text{cov}\{X, V\} - \omega^2 \, \text{var}\{X\} + \text{var}\{V\}. \tag{9.1.16}$$

The three ordinary differential equations (9.1.14) through (9.1.16) are readily decoupled and solved. In particular, we use (9.1.15) to eliminate $\text{cov}\{X, V\}$ and, subsequently, (9.1.16) to eliminate $\text{var}\{V\}$. This procedure generates the equation

$$\frac{d^3 \, \text{var}\{X\}}{dt^3} + 3\gamma \frac{d^2 \, \text{var}\{X\}}{dt^2} + (4\omega^2 + 2\gamma^2)\frac{d \, \text{var}\{X\}}{dt}$$

$$+ 4\gamma\omega^2 \left(\text{var}\{X\} - \frac{\beta^2}{2\gamma\omega^2} \right) = 0, \tag{9.1.17}$$

which can be expressed in terms of the auxiliary variable

$$y = \text{var}\{X\} - \frac{\beta^2}{2\gamma\omega^2} \tag{9.1.18}$$

in the convenient form

$$\frac{d^3 y}{dt^3} + 3\gamma \frac{d^2 y}{dt^2} + (4\omega^2 + 2\gamma^2)\frac{dy}{dt} + 4\gamma\omega^2 y = 0. \tag{9.1.19}$$

Equation (9.1.19) has solutions of form e^{pt}, where the constant p solves the indicial equation

$$p^3 + 3\gamma p^2 + (4\omega^2 + 2\gamma^2)p + 4\gamma\omega^2 = 0. \tag{9.1.20}$$

In general, cubic equations have complicated solutions. Fortunately, this cubic has relatively simple ones,

$$p = -\gamma, \, -\gamma \pm 2i\omega', \tag{9.1.21}$$

for which we recall that $\omega' = \sqrt{\omega^2 - \gamma^2/4}$. Therefore, the general solution of (9.1.17) is the linear combination

$$\text{var}\{X\} = \frac{\beta^2}{2\gamma\omega^2} + e^{-\gamma t}[a + be^{2i\omega' t} + ce^{-2i\omega' t}]. \tag{9.1.22}$$

The constants a, b, and c in (9.1.22) are determined by imposing the initial conditions $X(0) = x_0$ and $V(0) = v_0$ or their equivalents on (9.1.22), (9.1.16), and (9.1.15). In particular, recall that the variances and covariances of sure variables always vanish. Thus, $\text{var}\{X(0)\} = 0$, $\text{var}\{V(0)\} = 0$, and $\text{cov}\{X(0), V(0)\} = 0$. The results are

$$\text{var}\{X\} = \frac{\beta^2}{2\gamma\omega^2} + e^{-\gamma t}\left(\frac{\beta^2}{8\gamma\omega'^2\omega^2}\right)$$

$$\times [-4\omega^2 + \gamma^2\cos(2\omega't) - 2\gamma\omega'\sin(2\omega't)], \quad (9.1.23)$$

$$\text{cov}\{X, V\} = e^{-\gamma t}\left(\frac{\beta^2}{4\omega'^2}\right)[1 - \cos(2\omega't)], \quad (9.1.24)$$

and

$$\text{var}\{V\} = \frac{\beta^2}{2\gamma} + e^{-\gamma t}\left(\frac{\beta^2}{8\gamma\omega'^2}\right)[-4\omega^2 + \gamma^2\cos(2\omega't)$$

$$+ 2\gamma\omega'\sin(2\omega't)]. \quad (9.1.25)$$

As one might expect, in the long time limit $\gamma t \to \infty$, $\text{var}\{V\} \to \beta^2/2\gamma$, and, if the oscillator approaches thermal equilibrium with its environment so that

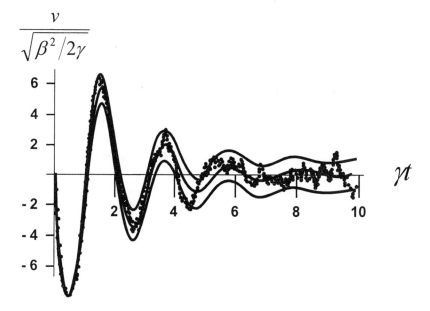

Figure 9.2. Dots represent the sample time evolution, while the solid lines represent the mean, and mean $\pm$ standard deviation of the normalized velocity coordinate $v/\sqrt{\beta^2/2\gamma}$ of a lightly damped ($\gamma = \omega/3$) stochastic harmonic oscillator.

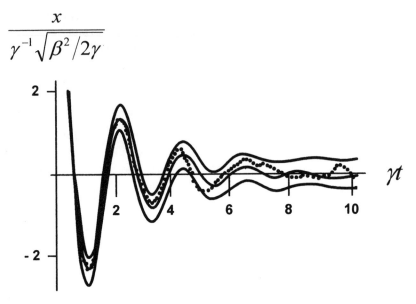

Figure 9.3. Dots represent the sample time evolution, while the solid lines represent the mean, and mean ± standard deviation of the normalized position coordinate $x/(\gamma^{-1}\sqrt{\beta^2/2\gamma})$ of a lightly damped ($\gamma = \omega/3$) stochastic harmonic oscillator.

var$\{V\} \to kT/M$, then, the fluctuation dissipation theorem, $\beta^2/2\gamma = kT/M$, also obtains. These expressions, (9.1.23) through (9.1.25), and those for mean$\{X(t)\}$ and mean$\{V(t)\}$, (9.1.9) and (9.1.10) respectively, completely describe the stochastic damped harmonic oscillator. See also Problem 9.3, *Oscillator Energy*.

Expressions (9.1.9), (9.1.10), and (9.1.23) through (9.1.25), pass the test of reducing to known results in appropriate limits. Of course, $t = 0$ recovers stipulated initial conditions, and $\beta^2 = 0$ reduces the random process to the familiar deterministic one. Less obvious is the $\omega^2 \to 0$ limit, which takes the oscillator process defined by (9.1.3) and (9.1.4) into Langevin's Brownian motion, defined by (8.1.4) and (8.1.5). See Problem 9.4, *O-U Process Limit*.

Figures 9.2 and 9.3 display sample speeds $V(t)$ and positions $X(t)$ of a lightly damped ($\gamma = \omega/3$) stochastic harmonic oscillator, as generated from equations (9.1.9) and (9.1.10) and (9.1.23) through (9.1.25) with the simulation method of section 8.2.

9.2 Stochastic Cyclotron Motion

One can add fluctuation and dissipation terms to any ordinary differential equation having time as an independent variable. Whether the resulting stochastic differential equation describes a physically meaningful process or not is, of course, another question. Adding dissipation and fluctuation to a charged particle in a magnetic field, in fact, makes perfect sense when the aim is to model simultaneous cyclotron motion and random scattering, as might occur, say, in the earth's magnetosphere, in a mass spectrometer, or in a particle accelerator.

A particle with charge Q and mass M in a stationary uniform magnetic field $\underline{B} = B\hat{z}$ obeys Newton's second law

$$M\frac{d\underline{v}}{dt} = Q\underline{v} \times \underline{B}, \tag{9.2.1}$$

or, equivalently,

$$d\underline{v} = (\underline{v} \times \underline{\Omega})dt, \tag{9.2.2}$$

where $\underline{\Omega} = \Omega\hat{z}$ and $\Omega(= QB/M)$ is the *cyclotron frequency*. Components of (9.2.2) in the $x - y$ plane, that is, in the plane normal to the magnetic field, are $dv_x = \Omega v_y dt$ and $dv_y = -\Omega v_x dt$. Adding dissipation and fluctuation to these produces the stochastic differential equations

$$dV_x = \Omega V_y dt - \gamma V_x dt + \sqrt{\beta^2 dt}\, N_{t,x}(0, 1) \tag{9.2.3}$$

and

$$dV_y = -\Omega V_x dt - \gamma V_y dt + \sqrt{\beta^2 dt}\, N_{t,y}(0, 1). \tag{9.2.4}$$

Here $N_{t,x}(0, 1)$ and $N_{t,y}(0, 1)$ are mutually independent and individually temporally uncorrelated unit normals. These, as well as

$$dX = V_x dt \tag{9.2.5}$$

and

$$dY = V_y dt, \tag{9.2.6}$$

govern the multivariate stochastic cyclotron process. A complete description of the particle dynamics also requires solving $dV_z = -\gamma V_z dt + \sqrt{\beta^2 dt}\, N_{t,z}(0, 1)$ and $dZ = V_z dt$ for motion parallel to the magnetic field. Since $V_z(t)$ is an O-U process and $Z(t)$ is its integral, we can refer to chapter 8 for their description.

The four coupled stochastic differential equations, (9.2.3) through (9.2.6), describing motion in the plane normal to the magnetic field, are linear in the dependent variables $X(t)$, $Y(t)$, $V_x(t)$, and $V_y(t)$. For this reason, these variables are different linear combinations of a single set of mutually independent unit normals and, therefore, via the normal sum theorem, are themselves correlated normals. Their complete description requires finding and solving the fourteen independent coupled ordinary differential equations governing the time dependence of their fourteen defining moments: mean$\{X\}$, mean$\{Y\}$, mean$\{V_x\}$, mean$\{V_y\}$, var$\{X\}$, var$\{Y\}$, var$\{V_x\}$, var$\{V_y\}$, cov$\{X, Y\}$, cov$\{X, V_x\}$, cov$\{X, V_y\}$, cov$\{Y, V_x\}$, cov$\{Y, V_y\}$, and cov$\{V_x, V_y\}$. This task is

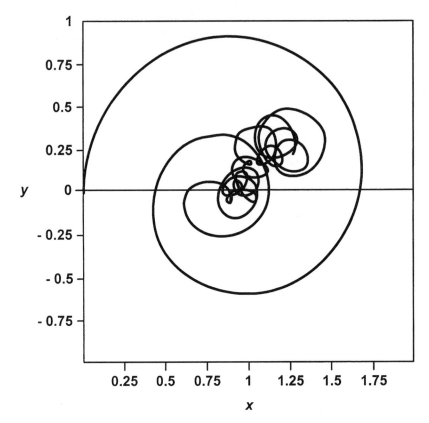

Figure 9.4. Trajectory of a stochastic magnetized charged particle described by the model of (9.2.3) through (9.2.6). See Lemons and Kaufman (1999).

too lengthy to outline here but requires only the methods already employed in chapter 8 and in section 9.1. Figure 9.4 shows a sample trajectory.

Here, instead of deriving the exact solutions necessary to generate particle trajectories like the one shown in figure 9.4, we use the Smoluchowski approximation to extract the physics of collision-limited charged particle diffusion across magnetic field lines. In particular, setting $dV_x = 0$ and $dV_y = 0$ in (9.2.3) and (9.2.4) and using (9.2.5) and (9.2.6) to eliminate velocity variables V_x and V_y, we have

$$0 = \Omega dY - \gamma dX + \sqrt{\beta^2 dt}\, N_{t,x}(0, 1) \qquad (9.2.7)$$

and

$$0 = -\Omega dX - \gamma dY + \sqrt{\beta^2 dt}\, N_{t,y}(0, 1). \qquad (9.2.8)$$

Multiplying each term in (9.2.7) by γ and each in (9.2.8) by Ω and adding the resulting equations produces

$$0 = -(\gamma^2 + \Omega^2)dX + \gamma\sqrt{\beta^2 dt}\, N_{t,x}(0, 1) + \Omega\sqrt{\beta^2 dt}\, N_{t,y}(0, 1), \qquad (9.2.9)$$

that is,

$$dX = \sqrt{\frac{\gamma^2\beta^2 dt}{(\gamma^2 + \Omega^2)^2}}\, N_{t,x}(0, 1) + \sqrt{\frac{\Omega^2\beta^2 dt}{(\gamma^2 + \Omega^2)^2}}\, N_{t,y}(0, 1). \qquad (9.2.10)$$

Similar manipulations eliminating X in favor of Y yield

$$dY = \sqrt{\frac{\gamma^2\beta^2 dt}{(\gamma^2 + \Omega^2)^2}}\, N_{t,y}(0, 1) - \sqrt{\frac{\Omega^2\beta^2 dt}{(\gamma^2 + \Omega^2)^2}}\, N_{t,x}(0, 1). \qquad (9.2.11)$$

Summing the two terms on the right-hand sides of (9.2.10) and (9.2.11) reveals that $X(t)$ and $Y(t)$ are Wiener processes with solutions

$$\text{mean}\{X\} = \text{mean}\{Y\} = \text{a constant} \qquad (9.2.12)$$

and

$$\text{var}\{X\} = \text{var}\{Y\} = \frac{\beta^2 t}{\Omega^2 + \gamma^2}. \qquad (9.2.13)$$

Thus the Smoluchowski approximation reduces stochastic cyclotron motion to cross-field diffusion with a diffusion constant $\frac{(\beta^2/2)}{\Omega^2+\gamma^2}$ reduced by a factor of $\gamma^2/(\Omega^2 + \gamma^2)$ from its field free ($\Omega^2 \to 0$) limit. Although not immediately obvious, the equations of motion (9.2.10) and (9.2.11) imply that X and Y are statistically independent. See Problem 9.5 *Statistical Independence*.

Problems

9.1. Smoluchowski Oscillator. Find the Smoluchowski approximation to the equations of motion (9.1.3) and (9.1.4) of a stochastic damped harmonic oscillator and solve for $X(t)$.

9.2. Critical Damping. Find mean$\{X\}$ and var$\{X\}$ for the critically damped ($\gamma = 2\omega$) stochastic oscillator by taking the appropriate limits of (9.1.9) and (9.1.23).

9.3. Oscillator Energy. The total energy of a simple harmonic oscillator is the sum of its kinetic $MV^2/2$ and potential $\omega^2 MX^2/2$ energies. Show that, in equilibrium when $\gamma t \to \infty$, solutions (9.1.23) and (9.1.25) and the fluctuation-dissipation theorem imply that the mean kinetic and potential energy are each equal to $kT/2$.

9.4. O-U Process Limit.   Show that the $\omega^2 \to 0$ limit reduces expressions (9.1.9) and (9.1.23) to those describing the mean and variance of the integral $X(t)$ of an O-U process as found in (8.1.18).

9.5. Statistical Independence.   Show that (9.2.10) and (9.2.11) imply that $\frac{d}{dt} \text{cov}\{X, Y\} = 0$, and, thus, given sure value initial conditions, that $\text{cov}\{X, Y\} = 0$ at all times.

10

Fluctuations without Dissipation

10.1 Effusion

Every physicist who maintains a vacuum system struggles to defeat *effusion*—the process whereby gas molecules flow through a small opening from one region (the environment) into another (the vacuum). Effusion can be quantified deterministically with a simple rate equation or stochastically as an O-U process. The latter, of course, includes the effect of fluctuations.

Figure 10.1 shows the situation we consider—a closed region divided into two compartments of volumes V_A and V_B containing, respectively, N_A and N_B molecules. An opening of area σ allows the molecules to move between compartments. We assume that the molecules are identical, that the gases are uniformly distributed within each compartment, that the gases are in thermal equilibrium with each other, and that the integers N_A and N_B are large enough to be treated as continuous variables. The rate at which molecules leave volume V_A must be proportional to their density N_A/V_A in compartment A. Likewise, the rate at which molecules enter volume V_A (from volume V_B) is proportional to N_B/V_B. Thus

$$\frac{dN_A}{dt} = -r\left(\frac{N_A}{V_A} - \frac{N_B}{V_B}\right), \qquad (10.1.1)$$

r is the effusion rate. The larger the area of the opening σ and the higher the gas temperature T, the more rapidly the molecules effuse; the larger the particle mass M, the more slowly the effusion. According to a simple model, $r = \sigma\sqrt{kT/(2\pi M)}$. By hypothesis, the total number of molecules is a constant $N_0 = N_A + N_B$. Using this relation to eliminate N_B from (10.1.1) transforms it into an equation for N_A alone,

$$\frac{dN_A}{dt} = -r\left(\frac{1}{V_A} + \frac{1}{V_B}\right)\left[N_A - \frac{N_0}{\left(\dfrac{V_B}{V_A} + 1\right)}\right]. \qquad (10.1.2)$$

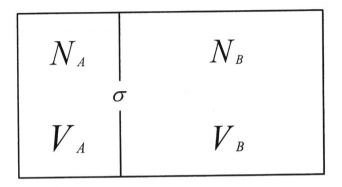

Figure 10.1. Effusion parameters.

Solving yields

$$N_A(t) = N_0 \left(\frac{V_A}{V_A + V_B} \right) + e^{-r\left(\frac{1}{V_A} + \frac{1}{V_B} \right)t} \left[N_A(0) - N_0 \left(\frac{V_A}{V_A + V_B} \right) \right].$$
(10.1.3)

As $t \to \infty$

$$N_A(t) \to N_0 \left(\frac{V_A}{V_A + V_B} \right),$$
(10.1.4)

and likewise

$$N_B(t) \to N_0 \left(\frac{V_B}{V_A + V_B} \right).$$
(10.1.5)

Thus, as $t \to \infty$ the densities in each compartment equalize.

More generally, N_A and N_B are random variables. In place of the deterministic rate equation (10.1.2) we propose the stochastic differential equation

$$dN_A = -\gamma(N_A - N_A^\infty)dt + \sqrt{\beta^2 dt}\, N_t(0, 1)$$
(10.1.6)

for N_A, where, for convenience we have adopted the notation $\gamma = r(\frac{1}{V_A} + \frac{1}{V_B})$ and $N_A^\infty = N_0(\frac{V_A}{V_A+V_B})$. Since (10.1.6) describes an O-U process, its solution is

$$N_A(t) = N \left(N_A^\infty + e^{-\gamma t}(N_A(0) - N_A^\infty), \frac{\beta^2}{2\gamma}(1 - e^{-2\gamma t}) \right).$$
(10.1.7)

However, var$\{N_A\}$ is not an energy, and therefore β^2 cannot be expressed in terms of a rate γ and a temperature T by requiring the equipartition of energy at equilibrium. The fluctuation-dissipation theorem does not apply; N_A and N_B fluctuate without dissipation.

Nonetheless, long-time, steady-state, or equilibrium values of mean$\{N_A\}$ and var$\{N_A\}$ provide a means of choosing the process parameters. From (10.1.7) we see that

$$\text{mean}\{N_A(\infty)\} = N_A^\infty \tag{10.1.8}$$

and

$$\text{var}\{N_A(\infty)\} = \frac{\beta^2}{2\gamma}. \tag{10.1.9}$$

Consider the following line of reasoning (also used in Problem 2.4, *Density Fluctuations*). A molecule must occupy a position in V_A or in V_B. Suppose these two mutually exclusive and exhaustive possibilities are realized with probabilities P_A and $P_B = 1 - P_A$. At equilibrium P_A and P_B are constant numbers. Thus mean$\{N_A(\infty)\}$ and var$\{N_A(\infty)\}$ must be functions of the equilibrium probabilities P_A and P_B. But what functions? To find out, let the random variables X_i with $i = 1, 2, \ldots N_0$ be a set of statistically independent indicator variables defined so that $X_i = 1$ when molecule i is within volume V_A and $X_i = 0$ when molecule i is within volume V_B. By design, the variables X_i characterize the gas in equilibrium. Clearly, $N_A = \sum_{i=1}^{N_0} X_i$, and, consequently,

$$\text{mean}\{X_i\} = 1 \cdot P_A + 0 \cdot P_B$$
$$= P_A \tag{10.1.10}$$

for all i. Likewise,

$$\text{var}\{X_i\} = (1 - \text{mean}\{X_i\})^2 \cdot P_A + (0 - \text{mean}\{X_i\})^2 \cdot P_B$$
$$= (1 - P_A)^2 \cdot P_A + (0 - P_A)^2 \cdot P_B$$
$$= (1 - P_A)P_A$$
$$= P_A P_B \tag{10.1.11}$$

for all i. Because the variables X_i are statistically independent,

$$\text{mean}\{N_A(\infty)\} = \sum_i^{N_0} \text{mean}\{X_i\}$$
$$= N_0 \text{mean}\{X_1\}$$
$$= N_0 P_A, \tag{10.1.12}$$

and, likewise,

$$\text{var}\{N_A(\infty)\} = N_0 P_A P_B. \qquad (10.1.13)$$

Combining these results with (10.1.8) and (10.1.9), we find that

$$N_A^\infty(\infty) = N_0 P_A \qquad (10.1.14)$$

and

$$\frac{\beta^2}{2\gamma} = N_0 P_A P_B. \qquad (10.1.15)$$

In terms of probabilities P_A and P_B, the O-U process (10.1.7) becomes

$$N_A(t) = N(N_0 P_A + e^{-\gamma t}(N_A(0) - N_0 P_A), N_0 P_A P_B(1 - e^{-2\gamma t})). \qquad (10.1.16)$$

We can reasonably assume that the equilibrium probability of being within a certain volume is proportional to that volume, or

$$P_A = V_A/(V_A + V_B) \qquad (10.1.17)$$

and

$$P_B = V_B/(V_A + V_B). \qquad (10.1.18)$$

Using (10.1.17) to eliminate P_A in (10.1.14) recovers (10.1.4), while using (10.1.17) and (10.1.18) to eliminate both P_A and P_B in (10.1.15) produces an expression for the characteristic fluctuation magnitude β^2.

This stochastic model of effusion extends the deterministic model, but both models have the same built-in artificiality: they ignore location within the compartments. For this reason, when either or both of the compartments are large, the assumption that equilibrium probabilities P_A and P_B are the same for each molecule becomes unrealistic. However, this problem doesn't arise when we apply the same mathematics to collections of quantum systems, as in Problem 10.1, *Two-Level Atoms*.

10.2 Elastic Scattering

A particle of mass M and speed V_0 moves among objects from which it scatters elastically, that is, without losing or gaining kinetic energy. Think, for instance, of electrons colliding with Helium ions in a hot plasma, neutrons diffusing through a matrix of cold material, elastically scattering photons, or even self-propelled flagellated bacteria swimming along at roughly constant speeds while randomly changing their heading (Berg 1993). Suppose that in

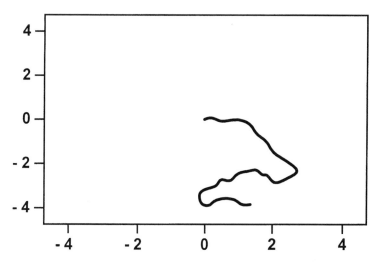

Figure 10.2. Direct numerical solutions of (10.2.2), (10.2.3), and (10.2.4), describing the trajectory of an elastically scattering particle. Axes are numbered in units of $V_0\gamma^{-1}$, and the total path length is $10\ V_0\gamma^{-1}$.

each small time interval dt the particle shifts its direction of propagation (in the x-y plane) only slightly by $d\Theta$ in such a way that

$$d\Theta = \sqrt{2\gamma\,dt}\,N_t(0, 1). \tag{10.2.1}$$

Thus $\Theta(t)$ is a Wiener process. Of course,

$$\Theta(t) = N(\Theta_0, 2\gamma t) \tag{10.2.2}$$

solves the stochastic differential equation (10.2.1). Here γ is a (positive definite) scattering rate having units of inverse time and so denoted, as we shall see, in order to emphasize the formal similarity with Brownian motion. Configuration-space coordinates X and Y are determined by a presumed constant speed V_0 and Θ through

$$dX = V_0 \cos \Theta dt \tag{10.2.3}$$

and

$$dY = V_0 \sin \Theta dt. \tag{10.2.4}$$

In principle, the time evolution of these three process variables Θ, X, and Y completely defines the time evolution of the elastically scattering particle.

Yet, when we attempt to solve (10.2.3) and (10.2.4) a difficulty arises. The trigonometric functions $\cos \Theta$ and $\sin \Theta$ are nonlinear in the random variable Θ and, for this reason, can't be integrated via the normal sum theorem. But we can eliminate the nonlinearity by replacing the random angle Θ with the random velocities

$$V_x = V_0 \cos \Theta \tag{10.2.5}$$

and

$$V_y = V_0 \sin \Theta. \tag{10.2.6}$$

The equation governing V_x comes from

$$
\begin{aligned}
dV_x &= d[V_0 \cos \Theta] \\
&= V_0[\cos(\Theta + d\Theta) - \cos \Theta] \\
&= V_0[\cos \Theta \cos d\Theta - \sin \Theta \sin d\Theta - \cos \Theta] \\
&= V_x[\cos d\Theta - 1] - V_y \sin d\Theta, \tag{10.2.7}
\end{aligned}
$$

which for small $d\Theta$ and $d\Theta = \sqrt{2\gamma \, dt} \, N_t(0, 1)$ becomes

$$
\begin{aligned}
dV_x &= -V_x \frac{(d\Theta)^2}{2} - V_y d\Theta \\
&= -V_x \gamma [N_t(0, 1)]^2 dt - V_y \sqrt{2\gamma \, dt} \, N_t(0, 1). \tag{10.2.8}
\end{aligned}
$$

The factor $[N_t(0, 1)]^2 dt$ requires special attention. Although in the form of a random variable, $[N_t(0, 1)]^2 dt$ can, without approximation, be replaced in (10.2.8) with the sure variable dt. The proof is simple. The moments of $[N_t(0, 1)]^2 dt$ and of dt are effectively identical through terms of order dt. In particular, $\langle [N_t(0, 1)]^2 dt \rangle = \langle dt \rangle = dt$ and $\langle ([N_t(0, 1)]^2 dt)^n \rangle \approx \langle dt^n \rangle = dt^n = 0$ for $n > 1$. Thus, the role played by $[N_t(0, 1)]^2 dt$ in the stochastic differential equation (10.2.8) is no different from the role played by dt. Consequently, equation (10.2.8) reduces to

$$dV_x = -\gamma V_x dt - V_y \sqrt{2\gamma \, dt} \, N_t(0, 1). \tag{10.2.9}$$

A similar derivation produces

$$dV_y = -\gamma V_y dt + V_x \sqrt{2\gamma \, dt} \, N_t(0, 1). \tag{10.2.10}$$

Note that the unit normals $N_t(0, 1)$ in the two equations (10.2.9) and (10.2.10) are completely correlated—in other words, they produce the same realizations. These equations of motion, along with

$$dX = V_x dt \qquad (10.2.11)$$

and

$$dY = V_y dt, \qquad (10.2.12)$$

govern the time evolution of this multivariate, but now linear, system.

The stochastic differential equations of motion (10.2.9) and (10.2.10) governing V_x and V_y have several noteworthy and useful properties. First, as expected, they exactly conserve kinetic energy $M(V_x^2 + V_y^2)/2$. To see this, use (10.2.9) to replace dV_x on the right-hand side of $dV_x^2 = 2V_x dV_x + (dV_x)^2$, discard terms smaller than dt, and replace $[N_t(0, 1)]^2 dt$ with dt. The result is

$$dV_x^2 = -2\gamma V_x^2 dt - 2V_x V_y \sqrt{2\gamma dt} N_t(0, 1) + 2\gamma V_y^2 dt \qquad (10.2.13)$$

and similarly

$$dV_y^2 = -2\gamma V_y^2 dt + 2V_y V_x \sqrt{2\gamma dt} N_t(0, 1) + 2\gamma V_x^2 dt. \qquad (10.2.14)$$

Adding these produces the expected statement of conservation

$$d(V_x^2 + V_y^2) = 0, \qquad (10.2.15)$$

whose solution is $V_x^2 + V_y^2 = V_0^2$. Second, the equations of motion (10.2.9) and (10.2.10) can be expressed compactly in vector form as

$$d\underline{V} = -\gamma \underline{V} dt - [\underline{V} \times \sqrt{2\gamma dt} N_t(0, 1)\hat{z}], \qquad (10.2.16)$$

where $\underline{V} = V_x \hat{x} + V_y \hat{y}$. Thus, the effect of elastic scattering is identical to the effect of a linear drag force $-\gamma M \underline{V}$ plus a fluctuating magnetic field

$$\underline{B} = -\frac{M}{Q} \sqrt{\frac{2\gamma}{dt}} N_t(0, 1)\hat{z} \qquad (10.2.17)$$

on a particle of charge Q and mass M. As $dt \to 0$, this fluctuating field becomes indefinitely large while its net effect over the interval dt becomes indefinitely small. Random variables with this behavior are said to exhibit *white noise*. Third, the structure of the fundamental equations of motion (10.2.9) through

(10.2.12) allows us to turn any valid expression derived from them into another valid expression by applying the transformation

$$V_x \rightarrow V_y, \qquad (10.2.18)$$

$$V_y \rightarrow -V_x, \qquad (10.2.19)$$

$$X \rightarrow Y, \qquad (10.2.20)$$

and

$$Y \rightarrow -X. \qquad (10.2.21)$$

Fourth, while the dynamical equations (10.2.9) and (10.2.10) are linear in the process variables V_x and V_y, they still contain $V_x N_t(0, 1)$ and $V_y N_t(0, 1)$, products of two random variables, one of which is a normal. Because such products are not normal variables, we still cannot use the normal sum theorem to solve (10.2.9) and (10.2.10). Neither can we exploit the central limit theorem— conservation of kinetic energy keeps the terms in (10.2.9) and (10.2.10) from being statistically independent. We can, however, derive ordinary differential equations that determine the time evolution of the moments of X, Y, V_x, and V_y. See Problem 10.3, *Mean Square Displacement*.

Taking the mean of equations (10.2.9) and (10.2.11) for V_x and X produces

$$\frac{d\langle V_x \rangle}{dt} = -\gamma \langle V_x \rangle \qquad (10.2.22)$$

and

$$\frac{d\langle X \rangle}{dt} = \langle V_x \rangle, \qquad (10.2.23)$$

which, given the sure initial conditions $X(0) = x_0$ and $V_x(0) = v_{x0}$, are solved by

$$\langle V_x \rangle = v_{x0} e^{-\gamma t} \qquad (10.2.24)$$

and

$$\langle X \rangle = x_0 + \frac{v_{x0}}{\gamma}(1 - e^{-\gamma t}). \qquad (10.2.25)$$

Applying the transformation (10.2.18) through (10.2.21) to these solutions yields $\langle V_y \rangle = v_{y0} e^{-\gamma t}$ and $\langle Y \rangle = y_0 + (v_{y0}/\gamma)(1 - e^{-\gamma t})$. Furthermore, from $dX^2 = 2X dX$ and (10.2.11) we find that $d\langle X^2 \rangle = 2\langle XV_x \rangle dt$, which, with (10.2.23), produces

$$\frac{d}{dt} \text{var}\{X\} = 2 \text{cov}\{X, V_x\}. \tag{10.2.26}$$

Likewise, from $d(XV_x) = X dV_x + V_x dX$ we find

$$\frac{d}{dt} \text{cov}\{X, V_x\} = \text{var}\{V_x\} - \gamma \text{cov}\{X, V_x\} \tag{10.2.27}$$

and, from $dV_x^2 = 2V_x dV_x + (dV_x)^2$,

$$\frac{d}{dt} \text{var}\{V_x\} = -2\gamma \text{var}\{V_x\} + 2\gamma \langle V_y^2 \rangle. \tag{10.2.28}$$

We can use conservation of energy (10.2.15) and the solution for $\langle V_x \rangle$ (10.2.24) to express the quantity $\langle V_y^2 \rangle$ in (10.2.28) in terms of $\text{var}\{X\}$, $\text{cov}\{X, V_x\}$, and $\text{var}\{V_x\}$ so that

$$
\begin{aligned}
\langle V_y^2 \rangle &= V_0^2 - \langle V_x^2 \rangle \\
&= V_0^2 - \text{var}\{V_x\} - \langle V_x \rangle^2 \\
&= V_0^2 - V_{x0}^2 e^{-2\gamma t} - \text{var}\{V_x\}.
\end{aligned} \tag{10.2.29}
$$

In this way, (10.2.28) becomes

$$\frac{d}{dt} \text{var}\{V_x\} = -4\gamma \text{var}\{V_x\} + 2\gamma (V_0^2 - V_{x0}^2 e^{-2\gamma t}). \tag{10.2.30}$$

Equations (10.2.26), (10.2.27), and (10.2.30), are coupled, linear, ordinary differential equations. Solving these for $\text{var}\{X\}$, we find

$$
\begin{aligned}
\text{var}\{X\} =& \left(\frac{V_0}{\gamma}\right)^2 \left[\gamma t + \frac{4e^{-\gamma t}}{3} - \frac{e^{-4\gamma t}}{12} - \frac{5}{4} \right] \\
&+ \left(\frac{V_{x0}}{\gamma}\right)^2 \left[\frac{e^{-4\gamma t}}{6} + \frac{4e^{-\gamma t}}{3} - e^{-2\gamma t} - \frac{1}{2} \right].
\end{aligned} \tag{10.2.31}
$$

Applying the transformation (10.2.18) through (10.2.21) to this solution generates a similar equation for $\text{var}\{Y\}$.

Apart from a drift velocity v_d, which can also be included in this calculation, the main difference between these results and those describing Langevin's

Brownian motion is the way in which the initial velocities V_{x0} and V_{y0} contribute (or do not contribute) to the time evolution of the spatial variances var$\{X\}$ and var$\{Y\}$. In the very long time regime ($\gamma t \gg 1$), elastic scattering reproduces, through leading order in $(\gamma t)^{-1}$, the linear time dependence characteristic of Brownian motion. That is,

$$\text{var}\{X\} \approx \text{var}\{Y\} \approx \frac{V_0^2 t}{\gamma}. \tag{10.2.32}$$

However, an expansion of (10.2.31) through leading order in γt yields

$$\text{var}\{X\} \approx \frac{2\gamma V_{y0} t^3}{3} \tag{10.2.33}$$

and, in like manner,

$$\text{var}\{Y\} \approx \frac{2\gamma V_{x0} t^3}{3}. \tag{10.2.34}$$

Apparently, at first, elastic scattering contributes more to the spatial variance in the plane normal to the initial velocity than in the direction of the initial velocity. In contrast, as shown in (8.1.17) the initial velocities do not appear at all in the expression for the spatial variance of a Brownian particle.

Problems

10.1. Two-Level Atoms.   A gas is composed of N_0 molecules, each one of which can occupy one of two states denoted A and B. In thermal equilibrium, the probability that a molecule occupies a state A is proportional to the Boltzmann factor, $e^{-E_A/kT}$, and the probability that it occupies state B is proportional to $e^{-E_B/kT}$, where E_A and E_B are allowed energy levels.

a. Find expressions for the equilibrium probabilities P_A and P_B in terms of E_A, E_B, and temperature T.
b. Given that the stochastic differential equation

$$dN_A = -\gamma_A(N_A - N_A^\infty)dt + \sqrt{\beta^2 dt}\,(0, 1)$$

governs the number of molecules in state A, evaluate the parameters N_A^∞ and β^2 in terms of γ_A, E_A, E_B, and T.

10.2. Cross-Field Diffusion.   Consider the ability of elastic scattering to cause the diffusion of a particle with charge Q and mass M across a stationary, uniform, magnetic field $\underline{B} = B_0\hat{z}$. In particular, add a Lorentz force to the equations of motion for elastic scattering (10.2.9) and (10.2.10), turning the latter into

$$dV_x = -\gamma V_x dt + V_y \Omega dt - V_y \sqrt{2\gamma dt} N_t(0, 1)$$

and

$$dV_y = -\gamma V_y dt - V_x \Omega dt + V_x \sqrt{2\gamma dt} N_t(0, 1)$$

where $\Omega_0 = QB/M$.

a. Show that these equations conserve kinetic energy, that is, show that $d(V_x^2 + V_y^2) = 0$.
b. Apply the Smoluchowski approximation—that is, set $dV_x = dV_y = 0$ and replace $V_x dt$ with dX and $V_y dt$ with dY. The Smoluchowski approximation extracts the physics of the limit in which configuration space diffusion is relatively large and velocity space diffusion is relatively small.
c. Separate dX and dY into two equations. Recall that the unit normal symbols $N_t(0, 1)$ appearing in the two equations denote the same variable.
d. Show that $\langle X \rangle = \langle Y \rangle = 0$.
e. Show that

$$\langle X^2 + Y^2 \rangle = \frac{2\gamma V_0^2 t}{[\Omega^2 + \gamma^2]}.$$

10.3. Mean Square Displacement.   Show that the mean square displacement of the elastically scattering particle with initial position $x_0 = 0$, $y_0 = 0$ is given by

$$\langle X^2 + Y^2 \rangle = \frac{2v_0^2}{\gamma^2}(\gamma t + e^{-\gamma t} - 1).$$

Appendix A

"On the Theory of Brownian Motion" by Paul Langevin

I. The very great theoretical importance presented by the phenomena of Brownian motion has been brought to our attention by Gouy.[1] We are indebted to this physicist for having clearly formulated the hypothesis that sees in the continual movement of particles suspended in a fluid an echo of molecular-thermal agitation and for having demonstrated this experimentally, at least in a qualitative manner, by showing the perfect permanence of Brownian motion and its indifference to external forces when the latter do not modify the temperature of the environment.

A quantitative verification of this theory has been made possible by Einstein,[2] who has recently given a formula that allows one to predict, at the end of a given time τ, the mean square $\overline{\Delta_x^2}$ displacement Δ_x of a spherical particle in a given direction x as the result of Brownian motion in a liquid as a function of the radius a of the particle, of the viscosity μ of the liquid, and of the absolute temperature T. This formula is

$$\overline{\Delta_x^2} = \frac{RT}{N} \frac{1}{3\pi \mu a} \tau, \tag{A.1}$$

where R is the ideal gas constant relative to one gram-molecule and N the number of molecules in one gram-molecule, a number well known today and around 8×10^{23}.

Smoluchowski[3] has attempted to approach the same problem with a method that is more direct than that used by Einstein in the two successive demonstrations he has given of his formula, and he has obtained for $\overline{\Delta_x^2}$ an expression of the same form as (1) but which differs from it by the coefficient 64/27.

II. I have been able to determine, first of all, that a correct application of the method of Smoluchowski leads one to recover the formula of Einstein *precisely*,

"Sur la théorie du mouvement brownien" *Comptes rendus Académie des Sciences* (Paris) *146*, (1908) 530–533. Translation by Anthony Gythiel, first published in *American Journal of Physics* *65* (1997): 1079–1081. Reprinted with permission. © 1997, American Association of Physics Teachers.

and, furthermore, that it is easy to give a demonstration that is infinitely more simple by means of a method that is entirely different.

The point of departure is the same: the theorem of the equipartition of the kinetic energy between the various degrees of freedom of a system in thermal equilibrium requires that a particle suspended in a liquid possesses, in the direction x, an average kinetic energy $RT/2N$ equal to that of a gas molecule of any sort, in a given direction, at the same temperature. If $\xi = dx/dt$ is the speed, at a given instant, of the particle in the direction that is considered, one therefore has, for the average extended to a large number of identical particles of mass m,

$$m\overline{\xi^2} = \frac{RT}{N}. \tag{A.2}$$

A particle such as the one we are considering, large relative to the average distance between the molecules of the liquid and moving with respect to the latter at the speed ξ, experiences (according to Stokes's formula), a viscous resistance equal to $-6\pi\mu a\xi$. In actual fact, this value is only a mean, and by reason of the irregularity of the impacts of the surrounding molecules, the action of the fluid on the particle oscillates around the preceding value, to the effect that the equation of the motion in the direction x is

$$m\frac{d^2x}{dt^2} = -6\pi\mu a\frac{dx}{dt} + X. \tag{A.3}$$

We know that the complementary force X is indifferently positive and negative and that its magnitude is such as to maintain the agitation of the particle, which, given the viscous resistance, would stop without it.

Equation (3), multiplied by x, may be written as:

$$\frac{m}{2}\frac{d^2x^2}{dt^2} - m\xi^2 = -3\pi\mu a\frac{dx^2}{dt} + Xx. \tag{A.4}$$

If we consider a large number of identical particles and take the mean of the equations (4) written for each one of them, the average value of the term Xx is evidently null by reason of the irregularity of the complementary forces X. It turns out that, by setting $z = \overline{dx^2/dt}$,

$$\frac{m}{2}\frac{dz}{dt} + 3\pi\mu az = \frac{RT}{N}.$$

The general solution

$$z = \frac{RT}{N}\frac{1}{3\pi\mu a} + Ce^{-\frac{6\pi\mu a}{m}t}$$

enters a *constant regime* in which it assumes the constant value of the first term at the end of a time of order $m/6\pi\mu a$ or of approximately 10^{-8} seconds for the particles for which Brownian motion is observable.

One therefore has, at a constant rate of agitation,

$$\frac{\overline{dx^2}}{dt} = \frac{RT}{N} \frac{1}{3\pi\mu a};$$

hence, for a time interval τ,

$$\overline{x^2} - \overline{x_0^2} = \frac{RT}{N} \frac{1}{3\pi\mu a}\tau.$$

The displacement Δ_x of a particle is given by

$$x = x_0 + \Delta_x,$$

and, since these displacements are indifferently positive and negative,

$$\overline{\Delta_x^2} = \overline{x^2} - \overline{x_0^2} = \frac{RT}{N} \frac{1}{3\pi\mu a}\tau;$$

thence the formula (1).

III. A first attempt at experimental verification has just been made by T. Svedberg,[4] the results of which differ from those given by formula (1) only by about the ratio 1 to 4 and are closer to the ones calculated with Smoluchowski's formula.

The two new demonstrations of Einstein's formula, one of which I obtained by following the direction begun by Smoluchowski, definitely rule out, it seems to me, the modification suggested by the latter.

Furthermore, the fact that Svedberg does not actually measure the quantity $\overline{\Delta_x^2}$ that appears in the formula and the uncertainty in the real diameter of the ultramicroscopic granules he observed call for new measurements. These, preferably, should be made on microscopic granules whose dimensions are easier to measure precisely and for which the application of the Stokes formula, which neglects the effects of the inertia of the liquid, is certainly more legitimate.

Notes

[1] Gouy, *Journ. de Phys.*, 2d ser., 7 (1888): 561; *Comptes rendus* 109 (1889): 102.

[2] A. Einstein, *Ann. d. Physik*, 4th ser., 17 (1905): 549; *Ann. d. Physik*, 4th ser., 19 (1906): 371.

[3] M. von Smoluchowski, *Ann. d. Physik*, 4th ser., 21 (1906): 756.

[4] T. Svedberg, *Studien zer Lehre von den kolloïden Lösungen* (Upsala, 1907).

Appendix B

Kinetic Equations

Chapter 6 presents two alternative but equivalent mathematical descriptions of a Wiener process: one in terms of the random variable $X(t)$ and its defining stochastic differential equation $dX = \sqrt{\delta^2\, dt}\, N_t(0, 1)$, and the other in terms of the probability density $p(x, t)$ and its defining partial differential equation

$$\frac{\partial p}{\partial t} = \left(\frac{\delta^2}{2}\right) \frac{\partial^2 p}{\partial x^2} .$$

All continuous, Markov, stochastic, normal processes have a similar dual description. Each two-variable process is governed by two stochastic differential equations of the general form

$$dV = a(X, V)\, dt + \sqrt{b^2(X, V)\, dt}\, N_t(0, 1) \tag{B.1}$$

and

$$dX = V\, dt, \tag{B.2}$$

where the functions $a(X, V)$ and $b(X, V)$ are general enough to accommodate many cases. What is the partial differential equation governing the equivalent two-variable probability density $p(x, v, t)$?

The key to converting between one description and the other is the identity

$$\iint f(x, v)\frac{\partial p}{\partial t}\, dx\, dv = \left\langle \frac{df(X, V)}{dt} \right\rangle, \tag{B.3}$$

where $f(X, V)$ is any smooth function of X and V. Now

$$
\begin{aligned}
df &= \frac{\partial f}{\partial X}\, dX + \frac{\partial f}{\partial V}\, dV + \frac{\partial^2 f}{\partial V^2}\frac{(dV)^2}{2} \\
&= \frac{\partial f}{\partial X} V\, dt + \frac{\partial f}{\partial V}\left[a\, dt + \sqrt{b^2\, dt}\, N_t(0, 1)\right] + \frac{\partial^2 f}{\partial V^2}\frac{b^2\, dt}{2}, \tag{B.4}
\end{aligned}
$$

where we have dropped terms smaller than dt. Substituting this result into (B.3) yields

$$\iint f\frac{\partial p}{\partial t}\, dx\, dv = \left\langle V\frac{\partial f}{\partial X} + a\frac{\partial f}{\partial V} + \frac{b^2}{2}\frac{\partial^2 f}{\partial V^2} \right\rangle \tag{B.5}$$

since $\langle\sqrt{b^2(X, V)}N_t(0, 1)\rangle = \langle\sqrt{b^2(X(t), V(t))}\rangle\langle N_t^{t+dt}(0, 1)\rangle = 0$. Expressing the right-hand side of (B.5) as an integration over phase space, we have

$$\iint f \frac{\partial p}{\partial t} \, dx \, dv = \iint \left[v \frac{\partial f}{\partial v} + a \frac{\partial f}{\partial x} + \frac{b^2}{2} \frac{\partial^2 f}{\partial v^2} \right] p \, dx \, dv, \qquad \text{(B.6)}$$

which, upon integrating the right-hand side by parts and dropping surface terms (at infinity), produces

$$\iint f(x, v) \frac{\partial p}{\partial t} \, dx \, dv$$

$$= \iint f(x, v) \left[-v \frac{\partial p}{\partial x} - \frac{\partial}{\partial v}(ap) + \frac{1}{2} \frac{\partial^2}{\partial v^2}(b^2 p) \right] dx \, dv. \qquad \text{(B.7)}$$

This equation holds for arbitrary function $f(x, v)$ if and only if

$$\frac{\partial p}{\partial t} + v \frac{\partial p}{\partial x} = -\frac{\partial}{\partial v}(ap) + \frac{1}{2} \frac{\partial^2}{\partial v^2}(b^2 p), \qquad \text{(B.8)}$$

which is the kinetic equation for arbitrary characterizing functions $a(x, v)$ and $b^2(x, v)$.

The O-U process stochastic differential equations are $dV = -\gamma V \, dt + \sqrt{\beta^2 \, dt} N_t(0, 1)$ and $dX = V \, dt$. Thus $a = -\gamma V$, $b^2 = \beta^2$, and the equivalent kinetic equation is the *Fokker-Planck equation*

$$\frac{\partial p}{\partial t} + v \frac{\partial p}{\partial x} = \gamma \frac{\partial}{\partial v}(vp) + \frac{\beta^2}{2} \frac{\partial^2 p}{\partial v^2}. \qquad \text{(B.9)}$$

The simple harmonic oscillator stochastic differential equations are $dV = -\omega^2 X \, dt - \gamma V \, dt + \sqrt{\beta^2 \, dt} N_t(0, 1)$ and $dX = V \, dt$. The equivalent kinetic equation,

$$\frac{\partial p}{\partial t} + v \frac{\partial p}{\partial x} = \frac{\partial}{\partial v}[(\omega^2 x + \gamma v)p] + \frac{\beta^2}{2} \frac{\partial^2 p}{\partial v^2}, \qquad \text{(B.10)}$$

is one example of the *Kramers kinetic equation*

$$\frac{\partial p}{\partial t} + v \frac{\partial p}{\partial x} = \frac{\partial}{\partial v} \left[\left(\frac{-F(x, v)}{m} + \gamma v \right) p \right] + \frac{\beta^2}{2} \frac{\partial^2 p}{\partial v^2}, \qquad \text{(B.11)}$$

describing the effect of an arbitrary smooth force $F(x, v)$ and a constant fluctuation parameter β^2 (Gardiner 1994).

Answers to Problems

Chapter 1

1.2   a. $\prod_{i=1}^{n} P_i$ and $1 - \prod_{i=1}^{n} P_i$

      b. $1 - \prod_{i=1}^{n}(1 - P_i)$

Chapter 2

2.1   mean$\{X\} = 3.50$, var$\{X\} = 2.92$, std$\{X\} = 1.71$, skewness$\{X\} = 0$, and kurtosis$\{X\} = 1.73$

2.3   a. mean$\{R\} = n\langle R_i \rangle$, var$\{R\} = n\left[\dfrac{t_i \langle R_i \rangle}{100}\right]^2$, $t = \dfrac{t_i}{\sqrt{n}}$

      b. mean$\{R\} = 50\Omega$, var$\{R\} = 10\Omega^2$, tolerance$\{R\} = 6.3\%$

2.4   a. mean$\{X_i\} = V/V_o$, var$\{X_i\} = (V/V_o)(1 - V/V_o)$

      b. mean$\{N\} = N_o(V/V_o)$, var$\{N\} = N_o(V/V_o)(1 - V/V_o)$, and

$$\sqrt{\text{var}\{N\}}/\text{mean}\{N\} = \sqrt{(1 - V/V_o)/(N_o V/V_o)}$$

Chapter 3

3.3   a. mean$\{X\} = n\mu$, var$\{X\} = n\sigma^2$, and $\langle X^2 \rangle = n\sigma^2 + n^2\mu^2$

      b. mean$\{X\} = (n/2)(\Delta x_r - \Delta x_l)$, var$\{X\} = \dfrac{n}{4}(\Delta x_r + \Delta x_l)^2$ and

$$\langle X^2 \rangle = \frac{n}{4}(\Delta x_r + \Delta x_l)^2 + \frac{n^2}{4}(\Delta x_r - \Delta x_l)^2$$

3.4   a. $m\Delta x^2$

      b. $\sqrt{m/n}$

3.5   a. mean$\{X_i\} = 1/2$, var$\{X_i\} = 1/4$

      b. mean$\{N\} = n/2$, var$\{N\} = N/4$

      c. mean$\{N/n\} = 1/2$, var$\{N/n\} = 1/(4n)$

Chapter 4

4.1   a. $p(x) = d/[\pi(x^2 + d^2)]$ for $-\infty < x < \infty$. Note that this probability distribution is that of a Cauchy variable $C(0, d)$.

4.3   b. $\lambda/(\lambda - t)$ for $t < \lambda$

       c. $1/\lambda^2$

       d. $n!/\lambda^n$.

4.4   a. 0.0361.

Chapter 5

5.3   a. ab

       b. $(a + b)^2 + c^2$

       c. $a^2 + b^2 + c^2$.

Chapter 6

6.1   a. $\delta^2 t'$

       b. $\sqrt{\frac{t'}{t}}$

6.2 $t = x_1^2/\delta^2$.

6.3   a. $X(t) = N_0^t(\alpha t, \delta^2 t)$.

       b. $p(x,t) = \dfrac{e^{-(x-\alpha t)^2/2\delta^2 t}}{\sqrt{2\pi \delta^2 t}}$

Chapter 7

7.1   a. $\gamma = F/(Mv_d)$

       b. $\beta^2 = 2FkT/(M^2 v_d)$.

Chapter 8

8.2 $1/\sqrt{2\gamma t}$.

8.3   a. $(8\beta^2 v_{y0}{}^3)/(3g^3)$

       b. $\sqrt{\dfrac{16\gamma^3 v_{y0}^3}{3g^3}}$

       c. 0.0007

Chapter 9

9.1 $X(t) = N\left(x_0 e^{-\omega^2 t/\gamma}, \left(\beta^2/2\gamma^3\right)\left(1 - e^{-2\omega^2 t/\gamma}\right)\right)$.

9.2 $\text{mean}\{X(t)\} = e^{-\omega t}(x_0 + v_0 t + \omega x_0 t)$,

     $\text{var}\{X(t)\} = \dfrac{\beta^2}{4\omega^3}[1 - e^{-2\omega t}(1 + 2\omega t + 2\omega^2 t^2)]$

Chapter 10

10.1   a. $P_A = e^{-E_A/kT} / \left(e^{-E_A/kT} + e^{-E_B/kT} \right),$

         $P_B = e^{-E_B/kT} / \left(e^{-E_A/kT} + e^{-E_B/kT} \right).$

     b. $N_A^{\infty} = N_0 e^{-E_A/kT} / \left(e^{-E_A/kT} + e^{-E_B/kT} \right),$

         $\beta^2 = 2\gamma_A N_0 e^{-(E_A+E_B)/kT} / \left(e^{-E_A/kT} + e^{-E_B/kT} \right)^2.$

References

Berg, Howard C. *Random Walks in Biology.* Princeton, N.J.: Princeton University Press, 1993. A brief introduction to the uses of random processes in biology that emphasizes the diffusion equation.

Brown, Robert. *Philosophical Magazine* 4 (1828): 161. Also in W. F. Magie, ed., *Source Book in Physics.* New York: McGraw-Hill, 1965. Pp. 251–55.

Bulmer, M. G. *Principles of Statistics.* New York: Dover, 1967. An excellent, compact introduction to probability and statistics.

Chandrasekhar, Subrahmanyan. "Stochastic Problems in Physics and Astronomy," *Reviews of Modern Physics*, January 1943. Reprinted in Nelson Wax, ed., *Selected Papers on Noise and Stochastic Processes*, New York: Dover, 1954. Many physicists have learned about stochastic processes from this quite demanding review article.

Coulston, C. C. *Dictonary of Scientific Biography.* New York: Scribner's, 1973. A standard multivolume biographical reference.

Courant, R., and H. Robbins. *What Is Mathematics?* New York: Oxford University Press, 1941. This book attempts to answer the question in its title by reproducing much important mathematical analysis. The result is an elegant and solid introduction to mathematics that has won the approval of two generations of readers and has remained in print for over sixty years.

Einstein, Albert. "Investigations on the Theory of Brownian Movement," *Annalen der Physik* (Leipzig) 17 (1905): 549–60. English translations appear in *Collected Papers of Albert Einstein*, 2:123–34, Anna Beck, trans. (Princeton: Princeton University Press, 1989), in *Albert Einstein, Investigations on the Theory of the Brownian Movement* (New York: Dover, 1956), 1–18, and in *Einstein's Miraculous Year* (Princeton: Princeton University Press, 1998). This is Einstein's most cited paper.

Fry, T. C. *Probability and Its Engineering Uses.* New York: Van Nostrand, 1928.

Gardiner, C. W. *Handbook of Stochastic Methods.* New York: Springer-Verlag, 1994. This book is a compendium of useful results, more an encyclopedia than a textbook.

Gillespie, Daniel. *Markov Processes: An Introduction for Physical Scientists.* New York: Academic Press, 1992. At a slightly more advanced level than the present book and substantially more comprehensive, Gillespie's book is my primary recommendation for further study.

———. "Mathematics of Brownian Motion and Johnson Noise," *American Journal of Physics* 64 (1996): 225–40.

Hacking, Ian. *The Emergence of Probability.* Cambridge: Cambridge University Press, 1975. A philosophical and historical analysis of various interpretations of the concept of probability.

Johnson, J. B. "Thermal Agitation of Electricity in Conductors," *Physical Review* 32 (1928): 97.

Kerrich, J. E. *An Experimental Introduction to the Theory of Probability.* Copenhagen: Munksgaard, 1946.

Klafter, J., M. F. Shlesinger, and G. Zumofen. "Beyond Brownian Motion," *Physics Today*, February 1996, 33. Describes the concept of a Levy process and its applications to "nonlinear, fractal chaotic, and turbulent systems."

——. "Above, below, and beyond Brownian Motion," *American Journal of Physics* 67 (1999): 1253–59.

Langevin, Paul. "On the Theory of Brownian Motion," *Comptes Rendues* 146 (1908): 530–533. An English translation of this paper, taken from D. S. Lemons and A. Gythiel, "Paul Langevin's 1908 Paper 'On the Theory of Brownian Motion'" *American Journal of Physics* 65 (1997): 1079–81, is reproduced, with permission, in Appendix A.

Laplace, Peirre Simon. *A Philosophical Essay on Probabilies*. New York: Dover, 1951. Originally published in 1820, this is an early, semipopular account of probability theory by the man who initiated the modern phase of its development.

Lemons, D. S., and D. L. Kaufman. "Brownian Motion of a Charged Particle in a Magnetic Field," *IEEE Transactions on Plasma Science* 27 (1999): 1288–96.

Springer, M. D. *The Algebra of Random Variables*. New York: Wiley, 1979. This is a specialist's monograph at a considerably higher level of difficulty than the present book.

Stigler, S. M. *The History of Statistics*. Cambridge, Mass.: Harvard University Press, 1986.

Uhlenbeck, G. E., and L. S. Ornstein. "On the Theory of Brownian Motion," *Physical Review* 36 (1930): 823–41. This paper also appears in Nelson Wax, ed., *Selected Papers on Noise and Stochastic Processes*. New York: Dover, 1954.

Index